CRM Short Courses

The volumes in the **CRM Short Courses** series have a primarily instructional aim, focusing on presenting topics of current interest to readers ranging from graduate students to experienced researchers in the mathematical sciences. Each text is aimed at bringing the reader to the forefront of research in a particular area or field, and can consist of one or several courses with a unified theme. The inclusion of exercises, while welcome, is not strictly required. Publications are largely but not exclusively, based on schools, instructional workshops and lecture series hosted by, or affiliated with, the *Centre de Researches Mathématiques* (CRM). Special emphasis is given to the quality of exposition and pedagogical value of each text.

More information about this series at http://www.springer.com/series/15360

The Virtual Series on Symplectic Geometry

Jun Zhang

Quantitative Tamarkin Theory

CENTRE
DE RECHERCHES
MATHÉMATIQUES

Springer

Jun Zhang
Département de Mathématiques
et Statistique
University of Montreal
Montréal, QC, Canada

ISSN 2522-5200 ISSN 2522-5219 (electronic)
CRM Short Courses
ISBN 978-3-030-37890-5 ISBN 978-3-030-37888-2 (eBook)
https://doi.org/10.1007/978-3-030-37888-2

Mathematics Subject Classification: 35A27, 37J10, 53D35, 55N99, 18A99

This Springer imprint is published by the registered company Springer Nature Switzerland AG.
The registered company address is: Gewerbestrasse 11, 6330 Cham, Switzerland

To my wife, Jue

Preface

The origin of this book goes back to the Fall of 2016 at Tel Aviv University, where a guided reading course was arranged by Leonid Polterovich, aiming at understanding the Guillermou-Kashiwara-Schapira sheaf quantization based on microlocal sheaf theory. This program was followed by two seminars at the Hebrew University of Jerusalem in the Fall of 2017 and at Tel Aviv University in the Spring of 2018, which attempted to understand how sheaf methods are applied as well as their relations to classical methods in symplectic geometry. This book grew out of these three seminars.

The microlocal sheaf theory developed by Kashiwara and Schapira in the '80s is a powerful theory which connects symplectic geometry, partial differential equations, and analysis. The pioneering work by Tamarkin in 2008 surprisingly illustrated how this theory can be used to solve non-displaceability problems, which constitutes a central subject in symplectic geometry. A series of works soon followed by different groups, devoted to translating more symplectic objects in the language of sheaves. In return, fruitful algebraic properties from sheaf theory became standard machinery applied in symplectic geometry and contact geometry. This book is an exposition of this fast developing and increasingly interesting subject. It focuses on the relations between symplectic geometry and Tamarkin category theory from a quantitative perspective, which includes Hofer's metric, barcodes, displacement energy, etc.

Thanks are due to many people. I want to thank Leonid Polterovich for providing me the opportunity to work in this field and for guiding me through some difficult concepts. I want to express my gratitude to those who attended my seminar course given in Spring 2018 at Tel Aviv University: Yaniv Ganor, Matthias Meiwes, Andrés Pedroza, Leonid Polterovich, Vukašin Stojisavljević, Igor Uljarevic and Frol Zapolsky. While writing this book, I got help from Semyon Alesker, Tomohiro Asano, Sheng-Fu Chiu, Leonid Polterovich and Nick Rozenblyum through many fruitful conversations, and I am grateful for their patience and inspiration.

The author was supported in part by the European Research Council Advanced Grant 338809.

Miami, FL, USA   Jun Zhang
July 2019

Contents

Chapter 1
Introduction

Abstract This chapter gives a brief overview of the primary materials in this book. It starts from the background of symplectic geometry with two famous results: Gromov's non-squeezing theorem and Arnold's conjecture (Lagrangian version). Then a discussion on the key concept of singular support follows, with an emphasis on its geometric interpretation. With the concept of singular support, Tamarkin categories will be described, and the Guillermou-Kashiwara-Schapira sheaf quantization will be formulated. These form the underlying platform where various symplectic objects can be expressed in terms of sheaves. Moreover, there is a section devoted to the background material on persistence **k**-modules, which can be viewed as elements in a special Tamarkin category; there is another section introducing Hofer's geometry, which is an iconic quantitative apparatus in symplectic geometry. Finally, a brief argument showing that the sheaf counterpart of the standard symplectic homology can be constructed from a certain projector in a Tamarkin category will be provided. This yield an alternative approach to study domains of Euclidean spaces.

1.1 A Brief Background of Symplectic Geometry

Symplectic geometry has its origin in the classical Hamiltonian mechanics in the 19th century. The total energy of a $2n$-dimensional mechanical system generates a flow, called *Hamiltonian flow*, which preserves the standard volume form of the phase space (this statement is called *Liouville's Theorem*). In fact, a stronger result holds. A Hamiltonian flow preserves a 2-form ω of the phase space, and the power ω^n is a volume form. In particular, ω satisfies two properties: it is closed, that is, $d\omega = 0$; and is non-degenerate, which precisely means that ω^n is a volume form. Any such 2-form is called a *symplectic form* or a *symplectic structure*. For instance, on $\mathbb{R}^{2n}$ with the coordinates $x_1, y_1, \ldots, x_n, y_n$, 2-form $\omega_{\mathrm{std}} := dx_1 \wedge dy_1 + \cdots + dx_n \wedge dy_n$ is the standard symplectic form on $\mathbb{R}^{2n}$. An even-dimensional manifold M^{2n} together with a symplectic form ω on it is called a *symplectic manifold*, denoted by a pair (M^{2n}, ω). Besides $(\mathbb{R}^{2n}, \omega_{\mathrm{std}})$, there are many other symplectic manifolds. Another often studied example is the

© Springer Nature Switzerland AG 2020
J. Zhang, *Quantitative Tamarkin Theory*, CRM Short Courses,
https://doi.org/10.1007/978-3-030-37888-2_1

cotangent bundle T^*M. There exists a canonical symplectic form $\omega_{\text{can}} = -d\theta_{\text{can}}$ θ_{can} is the canonical 1-form defined locally by $(\theta_{\text{can}})_{(q,p)}(v) = p(\pi_* v)$ at a point $(q, p) \in T^*M$. Here $\pi : T^*M \to M$ is the projection. Symplectic geometry studies those diffeomorphisms which preserve symplectic structures, i.e., diffeomorphisms $\phi : (M_1, \omega_1) \to (M_2, \omega_2)$ such that $\phi^* \omega_2 = \omega_1$. They are called *symplectic diffeomorphisms*. For instance, every Hamiltonian flow consists of symplectic diffeomorphisms. The time-1 map of a Hamiltonian flow is called a *Hamiltonian diffeomorphism*.

Denote by $\text{Ham}(M, \omega)$ the group of Hamiltonian diffeomorphisms, by $\text{Symp}(M, \omega)$ the group of symplectic diffeomorphisms, and by $\text{Diff}_{\text{vol}}(M)$ the group of volume preserving diffeomorphisms. In general, one has the following relations,

$$\text{Ham}(M, \omega) \subsetneqq \text{Symp}(M, \omega) \subsetneqq \text{Diff}_{\text{vol}}(M).$$

Both strict inclusions come from deep results in symplectic geometry, which establishes symplectic geometry as a subject worth of further study. Explicitly, $\text{Ham}(M, \omega) \subsetneqq \text{Symp}(M, \omega)$ (more precisely, $\text{Ham}(M, \omega)$ is contained in $\text{Symp}_0(M, \omega)$, the identity component) comes from the C^∞-*flux conjecture*. Roughly speaking, their difference is measured by $H^1(M; \mathbb{R})$. For more details, see Section 10.2 in [36]. The strict inclusion $\text{Symp}(M, \omega) \subsetneqq \text{Diff}_{\text{vol}}(M)$ was first observed by a celebrated result from Gromov [22] (called *Gromov's non-squeezing theorem*). It says that there is *no* symplectic embedding from a symplectic ball $B^{2n}(R)$ to a symplectic cylinder $Z^{2n}(r)$ if $R > r$. Here $B^{2n}(R) = \{(x_1, \ldots, y_n) \mid x_1^2 + \cdots + y_n^2 < R^2\}$ and $Z^{2n}(r) = \{(x_1, \ldots, y_n) \mid x_1^2 + y_1^2 < r^2\}$. In sharp contrast to this symplectic phenomenon, there always exists a "squeezing" by some $\phi \in \text{Diff}_{\text{vol}}(M)$ from $B^{2n}(R)$ into $Z^{2n}(r)$ even if R is much larger than r. This non-squeezing theorem motivates a fruitful research direction on symplectic embeddings.

Submanifolds of a symplectic manifold (M^{2n}, ω) are also of great interest. The first that come to mind are the *Lagrangian submanifolds*, say L, which are defined by the conditions $\omega|_L = 0$ and $\dim L = n$ (half of the dimension of the ambient symplectic manifold). For instance, the zero-section 0_M or in general any graph(df) of a differentiable function $f : M \to \mathbb{R}$ is a Lagrangian submanifold of $(T^*M, \omega_{\text{can}})$. Lagrangian submanifolds are important since they often show some rigidity in terms of intersection with each other. One famous result is the *Arnold conjecture* which says that $\phi(0_M) \cap 0_M \neq 0$ for any Hamiltonian diffeomorphism ϕ on T^*M. Moreover, under a transversality assumption on the intersections between $\phi(0_M)$ and 0_M, one can even provide a lower bound of the number of this intersection which can be as large as the sum of Betti numbers $\sum b_j(M; \mathbf{k})$. A different version of the Arnold conjecture (linked via the Weinstein neighborhood theorem by identifying the zero section 0_M with the diagonal of $M \times M$) is stated in terms of Hamiltonian diffeomorphisms. Explicitly, for any (non-degenerate) Hamiltonian diffeomorphism ϕ on a symplectic manifold (M, ω), the set of its fixed points $\text{Fix}(\phi)$ satisfies $\#\text{Fix}(\phi) \geq \sum b_j(M; \mathbf{k})$. This conjecture was proved by various people and groups such as Floer, Hofer-Salamon, Fukaya-Ono, Liu-Tian, Pardon. This is such a landmark result that many people think that the *modern* symplectic geometry emerged from this conjecture. Its proof also motivated various mathematical machinery developed particularly for symplectic geometry.

1.2 New Methods in Symplectic Geometry

Along with the work for proving Gromov's non-squeezing theorem and the Arnold conjecture, new methods were invented. Introduced by Gromov, the concept *J-holomorphic curve* is not only the key to proving his non-squeezing theorem, but also serves as a useful tool to create powerful invariants like the Gromov-Witten invariant (see Chapter 7 in [37]). Moreover, this directly inspired the Floer theory introduced by Floer in his original attempt to prove the Arnold conjecture. Nowadays, in various flavors, Floer theory has become one of the central theories in symplectic geometry. Roughly speaking, one can regard (the Hamiltonian) Floer theory as an ∞-dimensional Morse theory on the loop space or some cover of the loop space. Compared with the standard Morse theory, extraies come for the need to work in an infinite-dimensional space. This requires some serious results from Gromov's work [22].

Around the same time as the birth of Floer theory, Gromov's non-squeezing theorem and the Arnold conjecture inspired another machinery involving *generating functions*, which is based on works by Chaperon, Lalonde-Sikorav, Sandon, Théret, Traynor, and Viterbo (see [49] for a detailed review of this theory). As mentioned earlier, $\text{graph}(df)$ is a Lagrangian submanifold of T^*M, and we say that the function f *defines* the Lagrangian $\text{graph}(df)$. Although not every Lagrangian submanifold $L \subset T^*M$ behaves as nicely as a graph (for instance, possibly immersed), the method of generating functions enables us to still use a function to define this L by sacrificing the complexity of the ambient space. Explicitly, we say that L has a generating function $F(m, \xi) : M \times \mathbb{R}^K \to \mathbb{R}$ if L can be written as

$$L = \left\{ \left(m, \frac{\partial F}{\partial m}(m, \xi) \right) \middle| \frac{\partial F}{\partial \xi}(m, \xi) = 0 \right\},$$

where $\mathbb{R}^K$ is an auxiliary space introduced in order to resolve the "non-graphic" part of L. Under a certain assumption on the behavior at infinity, such F (if it exists) is uniquely defined up to stabilization by a quadratic form and composition with a fibered diffeomorphism of $M \times \mathbb{R}^K$. One standard example of a Lagrangian L which admits a generating function is one that is Hamiltonian isotopic to 0_M, i.e., $\phi(0_M)$ for some $\phi \in \text{Ham}(T^*M, \omega_{\text{can}})$. Note that critical points of the generating function F correspond to the intersection points of L with 0_M. Hence, this method brings intersection problems back to the classical Morse theory.

It is worth spelling out how one can associate a generating function to $\phi(0_M)$. Section 4 in [54] provides a detailed construction when $M = \mathbb{R}^{2n}$. First, break the Hamiltonian diffeomorphism ϕ into small pieces, that is,

$$\phi = \phi^{(N)} \circ \phi^{(N-1)} \circ \cdots \circ \phi^{(1)} \tag{1.1}$$

where each $\phi^{(i)}$ is a C^1-small Hamiltonian diffeomorphism. One can easily associate a generating function for each such C^1-small Hamiltonian diffeomorphism. Then by a composition formula (called Chekanov's formula, see [7]), we are able to "glue" all these pieces inductively to obtain a generating function for the given ϕ. This composition formula is crucial, but unfortunately appears to be quite complicated (see Section 8 in [49] for a geometric interpretation). Later we will see that the idea of this construction, in particular the composition formula, has an easy reformulation in the language of sheaves. This enables us to view symplectic geometry from another perspective.

1.3   Singular Support and Its Geometry

The concept *singular support* appears in different branches of mathematics, for instance, Fourier analysis (where it is called *wavefront set*), $\mathcal{D}$-modules (where it is called *characteristic variety*), and microlocal analysis in the framework of sheaf theory (or simply called microlocal sheaf theory) founded by Sato and by Kashiwara and Schapira. Interestingly enough, though in different areas, they are defined in a rather similar way and more or less related to each other. The singular support, denoted here by SS for brevity, serves as a useful tool that can transfer algebra to geometry.

Explicitly, for any $\mathcal{F} \in \mathcal{D}(\mathbf{k}_X)$, the derived category of sheaves of $\mathbf{k}$-modules over the manifold X, $SS(\mathcal{F})$, is a closed conical subset of the cotangent bundle T^*X. In particular, when $\mathcal{F}$ is constructible, $SS(\mathcal{F})$ is always a (singular) Lagrangian submanifold. For instance, take the sheaf $\mathbf{k}_{\{f(m)+t\geq 0\}}$ in the derived category $\mathcal{D}(\mathbf{k}_{M\times\mathbb{R}})$, where f is a differentiable function on M and t denotes the variable of the $\mathbb{R}$-component. One has the transform

$$\text{(function) } f \longrightarrow \text{(algebra) } \mathbf{k}_{\{f(m)+t\geq 0\}} \xrightarrow{\;SS + \text{reduction}\;} \text{(geometry) graph}(df), \tag{1.2}$$

where *reduction* is from $T^*(M \times \mathbb{R})$ to T^*M, simply by first intersecting with $\tau = 1$ and then projecting. Here, τ is the dual variable of t. With this procedure, one recovers the Lagrangian submanifold graph$(df) \subset T^*M$. Moreover, it is not hard to perform a similar process for a Lagrangian submanifold which admits a generating function. Then the interaction between Lagrangian submanifolds of T^*M can be transferred to an interaction between sheaves.

Back to the definition of SS, up to a closure, it measures the co-directions in T^*X where $\mathcal{F}$ *"does not propagate"*. Explicitly, we say the sheaf $\mathcal{F}$ does not propagate at (x, ξ) if there exists a smooth function $\phi : X \to \mathbb{R}$ satisfying $\phi(x) = 0$ and $d\phi(x) = \xi$, such that some section $H^*(\{\phi < 0\}; \mathcal{F})$ cannot be extended to a neighborhood of (x, ξ). For instance, if $\mathcal{F}$ is a constant sheaf, then $SS(\mathcal{F}) = 0_M$. Moreover, $SS(\mathcal{F}) \cap 0_M = \text{supp}(\mathcal{F})$, the support of $\mathcal{F}$. Therefore, to some extent, SS is a microlocal generalization of $\text{supp}(\mathcal{F})$. The definition of singular support has many interesting and meaningful consequences. On the one hand, directly from its definition, SS is rather computable in many elementary cases, see Subsect. 2.8; on the other hand, there is a rich pool of operators on sheaves such as Rf_*, f^{-1}, $\otimes$, $R\text{Hom}$, etc., which results in rich functorial behaviors of SS, see Subsect. 2.9. More interestingly, some fancier operators on sheaves (which are combinations of basic operators) correspond to well-known operators on Lagrangian submanifolds. Here are some examples.

operators	sheaves	Lagrangians
Corollary 2.1	external product $\mathcal{F} \boxtimes \mathcal{G}$	Cartesian product
Definition 3.4	composition $\mathcal{F} \circ \mathcal{G}$	Lag. correspondence
Definition 3.3	convolution $\mathcal{F} * \mathcal{G}$	fiberwise summation
Definition 3.10	"dual" convolution $\mathcal{H}om^*(\mathcal{F}, \mathcal{G})$	fiberwise subtraction

Finally, we emphasize that constraints on the singular support $SS(\mathcal{F})$ can force the sheaf $\mathcal{F}$ to behave in a rather restricted way. For instance, if $SS(\mathcal{F}) \subset 0_M$, then $\mathcal{F}$ has to be locally constant. The proof of this is more non-trivial than its statement, and requires a powerful result called *microlocal Morse theory*; see Theorem 2.10 in Subsect. 2.9.2.

1.4 Different Appearances of Tamarkin Category

So far, we have seen some stories in symplectic geometry (in particular on T^*M), as well as a possible way to associate sheaves to submanifolds in T^*M via generating functions. In general, one expects to tell as many stories in symplectic geometry (on T^*M) as possible in the language of sheaves. The first and foremost question is certainly the platform that we shall work on. Instead of $\mathcal{D}(\mathbf{k}_M)$ itself, we will start from $\mathcal{D}(\mathbf{k}_{M \times \mathbb{R}})$ as in the transform (1.2), where the extra $\mathbb{R}$-component plays the role of the conical property of the singular support. This conical property is, however, slightly unnatural from the perspective of the classical symplectic geometry. Besides this formal compatibility with the singular support, adding this extra variable $\mathbb{R}$ admits several meaningful explanations, and it turns out to be absolutely crucial in this book. In this section, we will emphasize the aspect that $\mathbb{R}$ can be regarded as a "ruler" for filtrations.

The *Tamarkin category* $\mathcal{T}(M)$ is defined as $\mathcal{D}_{\{\tau \leq 0\}}(\mathbf{k}_{M \times \mathbb{R}})^{\perp,l}$, the left orthogonal complement of the full triangulated subcategory $\mathcal{D}_{\{\tau \leq 0\}}(\mathbf{k}_{M \times \mathbb{R}})$ of $\mathcal{D}(\mathbf{k}_{M \times \mathbb{R}})$, where $\mathcal{D}_{\{\tau \leq 0\}}(\mathbf{k}_{M \times \mathbb{R}})$ consists of all the elements in $\mathcal{D}(\mathbf{k}_{M \times \mathbb{R}})$ such that their singular supports lie in $T^*M \times (\mathbb{R} \times \{\tau \leq 0\})$. Though it looks a little bizarre at a first glance, it is in fact an ingenious observation from [52] that every $\mathcal{F} \in \mathcal{D}(\mathbf{k}_{M \times \mathbb{R}})$ can be decomposed and fit into a distinguished triangle in $\mathcal{D}(\mathbf{k}_{M \times \mathbb{R}})$, that is,

$$\mathcal{F} * \mathbf{k}_{M \times [0,\infty)} \longrightarrow \mathcal{F} \longrightarrow \mathcal{F} * \mathbf{k}_{M \times (0,\infty)}[1] \xrightarrow{+1} . \qquad (1.3)$$

Here "$*$" is the sheaf convolution operator, $\mathcal{F} * \mathbf{k}_{M \times [0,\infty)} \in \mathcal{T}(M)$ and $\mathcal{F} * \mathbf{k}_{M \times (0,\infty)}[1] \in \mathcal{D}_{\{\tau \leq 0\}}(\mathbf{k}_{M \times \mathbb{R}})$. In fact, by orthogonality, $\mathcal{F} \in \mathcal{T}(M)$ if and only if $\mathcal{F} = \mathcal{F} * \mathbf{k}_{M \times [0,\infty)}$ (see Theorem 3.1 in Subsect. 3.3.2). This is a complete characterization of the elements in $\mathcal{T}(M)$, and it is exactly this characterization that highlights the role of $\mathbb{R}$ and the necessity of taking $\mathcal{T}(M)$ as our working platform instead of $\mathcal{D}(\mathbf{k}_{M \times \mathbb{R}})$. Explicitly, it's easy to see that shifting along $\mathbb{R}$, $(m, t) \to (m, t + c)$, induces a natural operator T_{c*} on any element in $\mathcal{D}(\mathbf{k}_{M \times \mathbb{R}})$. Furthermore, in $\mathcal{T}(M)$ for every pair $a \leq b$ there exists a well-defined functor $\tau_{a,b}$: $T_{a*}\mathcal{F} \to T_{b*}\mathcal{F}$ (which does *not* exist in $\mathcal{D}(\mathbf{k}_{M \times \mathbb{R}})$ in general!). Geometrically, people should regard this shift functor acting on $\mathcal{T}(M)$ as the change of levels of the sublevel sets in the classical Morse theory.

There are other Tamarkin categories with extra restrictions of singular supports from closed subsets $A \subset T^*M$. Denote by $\mathcal{T}_A(M)$ the full triangulated subcategory of $\mathcal{T}(M)$ that consists of sheaves $\mathcal{F}$ such that the reduction of $SS(\mathcal{F})$ lies in A. There are two types of subsets A we will mainly consider: A is a Lagrangian submanifold of T^*M, for instance $A = 0_M$ or its Hamiltonian deformations; or A is a closed (possibly unbounded) domain of T^*M, for instance $A = B(r)^c$, the complement of an open symplectic ball in $T^*\mathbb{R}^n (\simeq \mathbb{R}^{2n})$. Later we will see that the first case is related with the Arnold conjecture, while the second is related with Gromov's non-squeezing theorem .

Finally, let us address the case when the restriction subset is open, denoted by U. Directly mimicking the definition above does *not* work, because $\mathcal{T}_U(M)$ is not a well-defined *triangulated* subcategory. One correct way to define it is to let $A = U^c$ and define $\mathcal{T}_U(M) := \mathcal{T}_A(M)^{\perp}$, the (left or right) orthogonal complement of $\mathcal{T}_A(M)$ in $\mathcal{T}(M)$. In fact, Chiu's work [10] carefully deals with the case when $M = \mathbb{R}^n$ and $U = B(r)$. One of his main results is another *orthogonal* decomposition in the same spirit as (1.3). Explicitly, for any $\mathcal{F} \in \mathcal{T}(\mathbb{R}^n)$, there exist $P_{B(r)}$ and $Q_{B(r)}$ in $\mathcal{D}(\mathbf{k}_{\mathbb{R}^n \times \mathbb{R}^n \times \mathbb{R}})$ and a decomposition

$$\mathcal{F} \bullet P_{B(r)} \longrightarrow \mathcal{F} \longrightarrow \mathcal{F} \bullet Q_{B(r)} \xrightarrow{+1} \qquad (1.4)$$

such that $\mathcal{F} \bullet P_{B(r)} \in \mathcal{T}_{B(r)}(\mathbb{R}^n)$ and $\mathcal{F} \bullet Q_{B(r)} \in \mathcal{T}_{B(r)^c}(\mathbb{R}^n)$. Here "$\bullet$" is a mixture of composition and convolution operators (see Definition 3.5) and $P_{B(r)}$ is called the *ball-projector*. We will elaborate on this more in Sect. 1.7 and Subsect. 4.4.

Therefore, similarly to (1.3), $\mathcal{F} \in \mathcal{T}_{B(r)}(\mathbb{R}^n)$ if and only if $\mathcal{F} \bullet P_{B(r)} = \mathcal{F}$, which is a complete characterization of the elements in $\mathcal{T}_{B(r)}(\mathbb{R}^n)$. This construction can be extended to a more general open domain $U \subset T^*\mathbb{R}^n$ and then yields a restricted Tamarkin category $\mathcal{T}_U(\mathbb{R}^n)$. The associated P_U is called the U-projector.

All in all, various Tamarkin categories provide nice working environments where one can handle some standard objects in symplectic geometry. One remark is that $\mathcal{T}(M)$ or $\mathcal{T}_A(M)$ contain rich information since one has the freedom to choose our preferred elements inside to work with. This is actually a meaningful observation and will be useful in Sect. 1.6.

1.5 Inspirations from Persistence k-Modules

Before translating more symplectic geometry objects into Tamarkin categories, we want to take a detour to a new algebraic structure, called a *persistence* k-*module*, which is a quick and elegant way to package big data. Explicitly, a persistence k-module V is described as $\{\{V_t\}_{t \in \mathbb{R}}, \iota_{s,t}\}_{s \le t}$ where each V_t is a finite-dimensional k-module, and for each $s \le t$ the structure maps $\iota_{s,t} : V_s \to V_t$ satisfy $\iota_{t,t} = \mathbb{1}_{V_t}$ and if $r \le s \le t$, then $\iota_{r,t} = \iota_{s,t} \circ \iota_{r,s}$. A standard algebraic example is an *interval-type* persistence k-module, denoted by $\mathbb{I}_{(a,b]}$, where $(\mathbb{I}_{(a,b]})_t = \mathbf{k}$ if and only if $t \in (a, b]$, and the structure maps are non-zero and the identity on $\mathbf{k}$ if and only if both s, t lie in $(a, b]$. These interval-type k-modules form the building blocks of the general persistence k-modules, in the sense that one has a decomposition theorem, called Normal Form Theorem, which asserts that

$$V = \bigoplus \mathbb{I}_{(a_j, b_j]}^{m_j}, \quad \text{where } m_j \text{ is the multiplicity of } \mathbb{I}_{(a_j, b_j]}, \tag{1.5}$$

and moreover this decomposition is unique (up to reordering). Therefore, for each V, one can associate a collection of intervals (exactly) using this decomposition, i.e., $V \mapsto \mathcal{B}(V)$. This $\mathcal{B}(V)$ is called the *barcode* of V.

One clear motivation for the introduction of persistence k-modules comes from classical Morse theory. For any Morse function $f : M \to \mathbb{R}$, set $V_t := H_*(\{f < t\}; \mathbf{k})$. Homologies of such sublevel sets can be put together to form a persistence k-module denoted by $V(f)$, where $\iota_{s,t}$ is induced by the inclusion $\{f < s\} \hookrightarrow \{f < t\}$. Interestingly, for two such Morse functions f and g, $V(f)$ and $V(g)$ are comparable by the following sandwich-type inclusions

$$\{f < t\} \subset \{g < t + c\} \subset \{f < t + 2c\},$$

where $c := \|f - g\|_{C^0}$ (and the similar symmetric inclusion holds). Note that a uniform shift of these level sets corresponds to a uniform parameter-shift of persistence k-module, that is, $(V[c])_t := V_{t+c}$, and all related morphisms can also be shifted in the same manner. In general, in the language of persistence

k-modules, we have symmetric sandwich relations between the shifted V and the shifted W. Explicitly, there exist morphisms $F: V \to W[\delta]$ and $G: W \to V[\delta]$ such that

$$G[\delta] \circ F = \Phi_V^{2\delta} \quad \text{and} \quad F[\delta] \circ G = \Phi_W^{2\delta}, \tag{1.6}$$

where $\Phi_V^{2\delta} : V \to V[2\delta]$ are the canonical structure maps of V (and similarly for $\Phi_W^{2\delta}$). If (1.6) is satisfied, then V and W are said to be δ-*interleaved*. For instance, $V(f)$ and $V(g)$ above are c-interleaved. This interleaving relation provides a quantitative way to compare (and also to define a distance between) two persistence **k**-modules. Interested readers can refer to an active research direction called *topological data analysis* for numerous practical applications of this theory. Meanwhile, based on the Hamiltonian Floer theory, [44] introduced this language in the study of symplectic geometry for the first time.

The spirit of persistence **k**-modules spreads around in this book. Let us emphasize two things. One is that persistence **k**-modules in the book are very often replaced by the constructible sheaves over $\mathbb{R}$, mainly due to the following fact that is similar to (1.5) (see Theorem 1.15 in [34]): any constructible sheaf $\mathcal{F}$ over $\mathbb{R}$ admits a decomposition $\mathcal{F} \simeq \bigoplus \mathbf{k}_{I_j}$ with a (locally finite) collection of intervals $\{I_j\}_{j \in J}$. Moreover, this decomposition is unique, therefore we can associate the sheaf $\mathcal{F}$ with a collection of intervals, denoted by $\mathcal{B}(\mathcal{F})$, called the *sheaf barcode* of $\mathcal{F}$. There is a certain equivalence between persistence **k**-modules and constructible sheaves over $\mathbb{R}$ with some constraint on their singular supports, and this is discussed in details in the appendix, Sect. A.1. The other is the interaction relation in (1.6). Recall that one crucial feature of a Tamarkin category is the possibility of "shifting" elements along the $\mathbb{R}$-direction. It is not hard to see how this can be used to define an interleaving type pseudo-distance between two elements in $\mathcal{T}(M)$ (see $d_{\mathcal{T}(M)}$ in Definition 3.11), which will be useful once the displacement energy is involved. We will explain this in the next section.

1.6   Sheaf Quantization and the Hofer Norm

The key quantitative ingredient in symplectic geometry is the Hofer norm, which is defined on every Hamiltonian diffeomorphism. Explicitly, for any $\phi \in \mathrm{Ham}(M, \omega)$,

$$\|\phi\|_{\mathrm{Hofer}} := \inf \left\{ \int_0^1 (\max_M H_t - \min_M H_t) dt \ \middle| \ \phi_H^1 = \phi \right\}. \tag{1.7}$$

It is indeed a norm, in particular, non-degenerate. In fact, it is highly non-trivial to show that $\| \cdot \|_{\mathrm{Hofer}}$ is non-degenerate. The proof uses hard machinery like J-holomorphic curves. A geometric way to describe the Hofer norm is via the *displacement energy*, that is, for a given (closed) subset A, define $e(A) :=$

$\inf\{\|\phi\|_{\text{Hofer}} \mid \phi(A) \cap A = \emptyset\}$. The geometry derived from this norm is called *Hofer geometry*, and it guided the development of symplectic geometry over the past few decades. Interestingly, on T^*M this non-degeneracy can be confirmed by a sheaf method (and then one recovers a result from [42]). This is done in [1], where it is shown that for every closed ball with non-empty interior, its displacement energy is strictly positive (see Corollary 4.3). Its success is essentially attributed to a transformation from $\phi \in \text{Ham}(T^*M, \omega_{\text{can}})$ to its sheaf counterpart, called the *Guillermou-Kashiwara-Schapira sheaf quantization* of a Hamiltonian diffeomorphism ϕ based on [26]. For brevity, we refer to this as the GKS sheaf quantization.

Explicitly, for every compactly support $\phi \in \text{Ham}(T^*M, \omega_{\text{can}})$, first homogenize it to be a homogeneous Hamiltonian diffeomorphism Φ on $T^*(M \times \mathbb{R})$ (to make it compatible with our working space in the Tamarkin category $\mathcal{T}(M)$). Then the main result in [26] implies that there exists a *unique* $\mathcal{K}(= \mathcal{K}_\phi) \in \mathcal{D}(\mathbf{k}_{M \times \mathbb{R} \times M \times \mathbb{R}})$ such that $SS(\mathcal{K}) \subset \text{graph}(\Phi) \cup 0_{M \times \mathbb{R} \times M \times \mathbb{R}}$ and $\mathcal{K}|_{\Delta_\mathbb{R}} = \Delta_M$. Here $\Delta_\mathbb{R}$ is the diagonal of $\mathbb{R} \times \mathbb{R}$ and Δ_M is the diagonal of $M \times M$. Again, we have a transformation from dynamics to algebra then to geometry via SS, i.e., ϕ (or Φ) $\to \mathcal{K}_\phi \to \text{graph}(\Phi)$ as in (1.2). Convolution with $\mathcal{K}_\phi$ is a well-defined operator on $\mathcal{T}(M)$, and the general philosophy is

$$\text{convolution with } \mathcal{K}_\phi \iff \text{geometric action by } \phi,$$

i.e., for any $\mathcal{F} \in \mathcal{T}_A(M)$, $\mathcal{K}_\phi \circ \mathcal{F} \in \mathcal{T}_{\phi(A)}(M)$. Remarkably, the quantitative measurement by the pseudo-distance $d_{\mathcal{T}(M)}$, mentioned above in the interleaving style, says that $d_{\mathcal{T}(M)}(\mathcal{F}, \mathcal{K}_\phi \circ \mathcal{F})) \leq \|\phi\|_{\text{Hofer}}$. After assigning a well-defined capacity c to sheaves (see Sect. 3.9 and Sect. 4.3), this estimation yields a sheaf version of *energy-capacity inequality*

$$c(\mathcal{F}) \leq e(A) \quad \text{for any } \mathcal{F} \in \mathcal{T}_A(M).$$

As observed at the end of Sect. 1.4, clever choices of a sheaf $\mathcal{F}$ from $\mathcal{T}_A(M)$ yield interesting consequences. For instance, a "torsion" sheaf (see Example 4.7) results in the desired positivity conclusion when A is a closed ball in T^*M. For another example, a "non-torsion" sheaf (see Example 4.6) proves the Arnold conjecture when $A = 0_M$. This entire procedure is similar to the classical argument in symplectic geometry involving the displacement energy, where capacities of A are usually constructed based on some hard machinery like Floer theory or generating functions (see Section 5.3 in [57]). Here, everything is expressed in the language of sheaves, and is presented in the framework of Tamarkin categories. We expect this new approach to handle more *singular* situations that are beyond the reach of classical methods, see Remark 4.7.

Finally, it is enlightening to review the construction of the GKS sheaf quantization which illuminates the advantage of using sheaves (for more details, see Sect. 4.1). Let us state their theorem first: for any homogeneous Hamiltonian isotopy

$\Phi = \{\phi_t\}_{t \in I}$ on $\dot{T}^*X$ (T^*X with the zero section 0_X removed), there exists a unique element $\mathcal{K} \in \mathcal{D}(\mathbf{k}_{I \times X \times X})$ such that

(i) $SS(\mathbf{k}) \subset \Lambda_\Phi \cup 0_{I \times X \times X}$;
(ii) $\mathcal{K}|_{t=0} = \mathbf{k}_\Delta$, where Δ is the diagonal of $X \times X$.

Here Λ_Φ is the time-involving trace of the negative of graph(ϕ_t), called the *Lagrangian suspension* of Φ. Explicitly, if H_t denotes the Hamiltonian function generating the isotopy Φ, then

$$\Lambda_\Phi := \left\{ (z, -\phi_t(z), t, -H_t(\phi_t(z))) \,\middle|\, z \in \dot{T}^*X \right\}.$$

This is the right geometric realization of a Hamiltonian isotopy (and graph(ϕ_1) is just the restriction of Λ_Φ at $t = 1$). The basic idea of the construction of $\mathcal{K}$ is very similar to the generating function theory above. There are two steps: break into "small" Hamiltonian diffeomorphisms or isotopies as in (1.1); then glue these together. Instead of gluing *generating functions*, which appears to be complicated, the GKS method is to glue *sheaves* that are associated to those C^1-small Hamiltonian diffeomorphisms or isotopies. The magic operator for this sheaf gluing is simply the sheaf convolution "$\circ$". Explicitly, $\mathcal{K}$ is constructed in the form

$$\mathcal{K} = \mathcal{K}_1 \circ \mathcal{K}_2 \circ \cdots \circ \mathcal{K}_N$$

for some "small" $\mathcal{K}_i$, where condition (i) above ensures that the resulting $\mathcal{K}$ can represent the correct geometry. Finally, let us remark that one cannot skip the role of isotopy, i.e., time I-component, in this construction if we want to prove the uniqueness of such $\mathcal{K}$. See Subsect. 4.1.3 for a detailed discussion, where such uniqueness is guaranteed by some constraints on singular supports. In this sense, the GKS's sheaf quantization has a strong microlocal flavor.

1.7 U-Projector and Symplectic Homology

Gromov's non-squeezing theorem is a story about (reasonable) open domains of $\mathbb{R}^{2n}$. A direct and more symplectic way to handle such a domain U is by using its symplectic homology, denoted $\mathrm{SH}_*(U)$. Although there are many versions of symplectic homology (see [40] for a survey comparing different such versions), each of them is built from a certain limit of Hamiltonian Floer theories. Sect. 4.8 gives an explanation of this construction for the symplectic ball $B^{2n}(r)$. Roughly speaking, it characterizes a domain using the dynamics of its boundary. Historically, one application of symplectic homology is to associate various symplectic invariants to U, for instance, some symplectic capacities $c(U)$ (see [29, 59] and [48]). In fact, the existence of any symplectic capacity is equivalent to Gromov's non-squeezing theorem. Moreover, $\mathrm{SH}_*(U)$ can also be viewed as a persistence $\mathbf{k}$-module, and any such symplectic capacity admits an "easy" explanation in terms of the corresponding barcode. Keeping in mind the equivalence between persistence **k**-modules and constructible sheaves over $\mathbb{R}$, it is natural to look for a sheaf counterpart of $SH_*(U)$. Interestingly, the U-projector $P_U \in \mathcal{D}(\mathbf{k}_{\mathbb{R}^n \times \mathbb{R}^n \times \mathbb{R}})$ that appeared in the construction/decomposition in $\mathcal{T}_U(\mathbb{R}^n)$ above can provide such a counterpart. Explicitly, our sheaf analog is

$$\mathcal{F}(U) := R\pi_! \Delta^{-1} P_U \in \mathcal{D}(\mathbf{k}_{\mathbb{R}}) \quad \text{where} \quad \mathbb{R} \xleftarrow{\pi} \mathbb{R}^n \times \mathbb{R} \xrightarrow{\Delta} \mathbb{R}^n \times \mathbb{R}^n \times \mathbb{R}. \quad (1.8)$$

Here Δ is the diagonal embedding from the $\mathbb{R}^n$-component to the $\mathbb{R}^n \times \mathbb{R}^n$-component, and π is the projection onto the $\mathbb{R}$-component. For instance, one can compute the sheaf barcode $\mathcal{B}(\mathcal{F}(B(r))) = \{[m\pi r^2, (m+1)\pi r^2)\}_{m \geq 0}$, and it coincides with the standard barcode of symplectic homology $SH_*(B(r))$.

Digging into the construction of P_U, such a coincidence is not surprising at all. Let us roughly unravel the formula of P_U. Label the time by variable a (where its co-variable is denoted by b) and label the extra $\mathbb{R}$-component by t. One defines (see (10) in [10])

$$P_U := \mathbf{k}_{\{S+t \geq 0\}} \bullet_{\mathbb{R}_a} \mathbf{k}_{\{t+ab \geq 0\}}[1] \circ_{\mathbb{R}_b} \mathbf{k}_{\{b < r^2\}}, \quad (1.9)$$

where S is a generating function of the Hamiltonian isotopy ϕ_a generated by (any) Hamiltonian function H that defines U in the sense that $U = \{H < 1\}$. The rigorous construction of P_U needs some additional convolutions in terms of variable a since S is not well-defined for all $a \in \mathbb{R}$ (see Subsect. 4.5.1).

Two remarks are in order. (i) By orthogonality in the decomposition (1.4), one can show that P_U is independent of any such defining Hamiltonian functions (see Sect. 4.6); (ii) carefully computing $SS(P_{B(r)})$, one shows that the operators "$\bullet_{\mathbb{R}_a}$" and "$\circ_{\mathbb{R}_b}$" in the construction (1.9) are deliberately designed so that P_U behaves like a projector in the sense that action $\mathcal{F} \bullet_{\mathbb{R}^n} P_U$ cuts the part of $\mathcal{F}$ that is outside U (see Subsect. 4.5.2 for a detailed explanation of the operator $\bullet_{\mathbb{R}^n} P_U$ in geometric terms[1]). People who are familiar with symplectic homology should be aware of the similarity between (i), (ii) here and the well-known features of symplectic homology that its computation is independent of the choice of admissible Hamiltonian functions. These admissible Hamiltonian functions can be chosen from those supported *inside U*.

The similarity between $SH_*(U)$ and $\mathcal{F}(U)$ can also be explained by a geometric meaning of the stalks of $\mathcal{F}(U)$ (see Lemma 4.5 for the case where $U = B(r)$). Roughly speaking, for any given filtration $\lambda \in \mathbb{R}$,

$$\mathcal{F}(U)_\lambda = H_c^*(\{\text{level set defined by } S \text{ and } \lambda\}; \mathbf{k})$$

---

[1] This is based on a discussion and joint work with Leonid Polterovich.

where S is a generating function of dynamics associated to U as above. The generators of this cohomology are some critical points of S. Hence, due to Δ^{-1} in (1.8), Hamiltonian *loops* and only loops with actions bounded by λ are counted. This should remind the reader of Traynor's fundamental work [54] that defines symplectic homology via generating functions.

To sum up, we have seen that P_U and $\mathcal{F}(U)$ concisely package all the necessary elements to build a cohomology theory over a given domain U. Based on many functorial properties that $\mathcal{F}(U)$ enjoys, Sect. 4.10 shows how they can easily imply Gromov's non-squeezing theorem.

1.8 Further Discussions

There are many interesting sheaf-symplectic-related topics that are not included in this book. For instance, a modification of our discussion based on the ball-projector can be used to prove contact the non-squeezing theorem [15]. This is the main result obtained in [10] (and also in [18] by a more symplectic approach). Different groups successfully apply sheaf methods to knot theory and some related homology theories [39, 51], etc. At the same time, there exists a relation between microlocal sheaf theory and the Fukaya category discovered by [38] and motivated by Kontsevich's mirror symmetry. From a different background, in [55] in the language of deformation quantization, another category is established which partially shares some common features with Tamarkin categories. It will be a very interesting research direction to determine how our quantitative perspective can fit into some of the aforementioned works. It is worth mentioning that Tamarkin category also has an influence on D'Agnolo and Kashiwara's recent work on a Riemann-Hilbert correspondence for holonomic $\mathcal{D}$-module [14]. This is a successful attempt to generalize the classical Riemann-Hilbert correspondence [31] to handle the holonomic $\mathcal{D}$-modules with irregular singularities. Last but not least, in this book all the Tamarkin category stories take place only in cotangent bundles (partially due to the fact that singular supports naturally lie in the cotangent bundles). It is a challenging question how to generalize them to more general symplectic manifolds so that this sheaf method can recover more classical symplectic geometry results, or even find something new. Tamarkin's work [53] proposed and developed a well-defined microlocal category over any compact prequantizable symplectic manifold. This definitely needs to be digested more thoroughly by the general public. Finally, parallel to this book (but at a more advanced level), the recent work from Guillermou [25] provides a detailed overview of microlocal sheaf theory and its applications to various topics in symplectic geometry.

Chapter 2
Preliminaries

Abstract In this chapter, we give detailed descriptions of the concepts of derived category, persistence **k**-module, and singular support. These serve as preparations for the topics treated in later chapters. Sections 2.1 to 2.5 are devoted to derived category and derived functors, as well as their applications in the category of sheaves[1]. These are basic ingredients for Tamarkin categories. Sections 2.6 and 2.7 are devoted to the theory of persistence **k**-module theory, which figures the interleaving distance and barcodes[2]. Two main theorems highlight this theory: Normal Form Theorem and Isometry Theorem. Sections 2.8 and 2.9 are devoted to the definition of the singular support, its various functorial properties, and an important result, the microlocal Morse lemma, which generalizes the classical Morse lemma to a microlocal formulation[3]. This lemma is essential in the constructions of Tamarkin categories.

2.1 i-th Derived Functor

In this section, we will define a family of functors called i-th derived functors, and show how these functors are constructed and computed. Let $\mathcal{A}$ be an abelian category, for instance, $\mathcal{A} = \mathrm{Sh}(X, \mathcal{G})$, the category of sheaves of groups, or $\mathcal{A} = \mathrm{Sh}(\mathbf{k}_X)$, the category of sheaves of **k**-modules over a topological space X, or $\mathcal{A} = \mathrm{Mod}_R$, the category of R-modules, where R is a commutative ring.

Definition 2.1 Define the *chain complex category* of an abelian category $\mathcal{A}$ as follows,

$$\mathrm{Com}(\mathcal{A}) = \left\{ X_{\bullet} = \cdots \longrightarrow X_{i-1} \xrightarrow{d_{i-1}} X_i \xrightarrow{d_i} X_{i+1} \longrightarrow \cdots \,\middle|\, \begin{matrix} X_i \in \mathcal{A}, \ \forall i \\ d_i \circ d_{i-1} = 0 \end{matrix} \right\}.$$

---

[1]These are based on lectures given by Yakov Varshavsky at the Hebrew University of Jerusalem in the Fall of 2017.

[2]These are based on lectures given by Leonid Polterovich at the Hebrew University of Jerusalem in the Fall of 2017.

[3]These are based on a joint work with Asaf Kislev at Tel Aviv University in the Fall of 2016.

© Springer Nature Switzerland AG 2020
J. Zhang, *Quantitative Tamarkin Theory*, CRM Short Courses,
https://doi.org/10.1007/978-3-030-37888-2_2

To each element $X_\bullet \in \mathrm{Com}(\mathcal{A})$, we can associate a $\mathbb{Z}$-family of elements in $\mathcal{A}$, namely,

$$h^i(X_\bullet) = \frac{\ker(d_i)}{\mathrm{Im}(d_{i-1})} \in \mathcal{A},$$

which is called the *i-th cohomology* of $X_\bullet$, for each $i \in \mathbb{Z}$. In particular, if $h^i(X_\bullet) = 0$ for each $i \in \mathbb{Z}$, then $X_\bullet$ is called *exact*. In particular, an exact sequence of the form of $0 \longrightarrow X \longrightarrow Y \longrightarrow Z \longrightarrow 0$ is called a *short exact sequence*.

Definition 2.2 Let $F : \mathcal{A} \longrightarrow \mathcal{B}$ be an additive functor. The functor F is called *exact* if for any short exact sequence $0 \longrightarrow X \longrightarrow Y \longrightarrow Z \longrightarrow 0$ in $\mathcal{A}$,

$$0 \longrightarrow F(X) \longrightarrow F(Y) \longrightarrow F(Z) \longrightarrow 0$$

is also a short exact sequence in $\mathcal{B}$.

Exercise 2.1 If F is an exact functor, then for any $X_\bullet \in Com(\mathcal{A})$, $h^i(F(X_\bullet)) = F(h^i(X_\bullet))$. In particular, for any exact sequence $X_\bullet \in Com(\mathcal{A})$, $F(X_\bullet)$ is also an exact sequence.

Remark 2.1 Pathetic reality: many functors are not exact! In order to deal with this situation, we will "embed" $\mathcal{A}$ into a bigger category $\mathcal{D}(\mathcal{A})$ (called the derived category of $\mathcal{A}$, defined in Sect. 2.2) such that for any functor $F : \mathcal{A} \longrightarrow \mathcal{B}$, one gets an upgraded functor $RF : \mathcal{D}(\mathcal{A}) \longrightarrow \mathcal{D}(\mathcal{B})$ that is "exact". See Theorem 2.4 and Remark 2.5 for a precise formulation of this procedure.

Some functors are *left exact*, that is, for any short exact sequence $0 \longrightarrow X \longrightarrow Y \longrightarrow Z \longrightarrow 0$, we get an exact sequence

$$0 \longrightarrow F(X) \longrightarrow F(Y) \longrightarrow F(Z).$$

Example 2.1 The following examples are left exact functors. For the sake of brevity, denote by $\mathrm{Sh}(X)$ or $\mathrm{Sh}(Y)$ the category of sheaves of abelian groups.

(1) Define the functor $\Gamma(X, \cdot) : \mathrm{Sh}(X) \longrightarrow \mathcal{G}$ by $\mathcal{F} \mapsto \Gamma(X, \mathcal{F})$, *taking global sections* of a sheaf. Similarly, $\Gamma_c(X, \cdot)$ is defined by taking the compactly supported global sections.

(2) Let $f : X \longrightarrow Y$ be a continuous map. Define the *pushforward (or direct image)* functor $f_* : \mathrm{Sh}(X) \longrightarrow \mathrm{Sh}(Y)$ by $(f_*\mathcal{F})(U) := \mathcal{F}(f^{-1}(U))$.

(3) Let $f : X \longrightarrow Y$ be a continuous map where X and Y are locally compact. Define the *proper pushforward (or direct image with compact support)* functor $f_! : \mathrm{Sh}(X) \longrightarrow \mathrm{Sh}(Y)$ by

$$(f_!\mathcal{F})(U) = \{s \in (f_*\mathcal{F})(U) \mid f : \mathrm{supp}(s) \longrightarrow U \text{ is proper}\}.$$

In particular, $f_!\mathcal{F}$ is a subsheaf of $f_*\mathcal{F}$.

Exercise 2.2 Prove that in Example 2.1 these functors are indeed left exact.

Now we will clarify the "exactness" of RF mentioned earlier by using the following theorem (see Theorem 1.1 A. and Corollary 1.4. in Section 1 of Chapter III in [28]) which claims that by adding extra terms we can complete the "half" exact sequence induced by a left exact functor into a long exact sequence.

Theorem 2.1 *Let $\mathcal{A}$ be an abelian category (satisfying some condition $(*)$ specified later), and $F : \mathcal{A} \longrightarrow \mathcal{B}$ be a left exact functor. Then there exists a sequence of functors $R^i F : \mathcal{A} \longrightarrow \mathcal{B}, i = 0, 1, \ldots,$ such that*

(1) *$R^0 F = F$;*
(2) *for any short exact sequence $0 \longrightarrow X \longrightarrow Y \longrightarrow Z \longrightarrow 0$, there exists a long exact sequence*

$$\cdots \longrightarrow R^i F(X) \longrightarrow R^i F(Y) \longrightarrow R^i F(Z) \xrightarrow{\delta_i}$$
$$R^{i+1} F(X) \longrightarrow R^{i+1} F(Y) \longrightarrow R^{i+1} F(Z) \longrightarrow \cdots$$

(3) *the long exact sequence in (2) is functorial in the sense that a morphism between two short exact sequences induces a morphism between the corresponding long exact sequences;*
(4) *it is universal among any such family of functors satisfying (1) - (3).*

Definition 2.3 Given a left exact functor F, the functor $R^i F$ generated according to Theorem 2.1 is called the *i-th derived functor of F*.

2.1.1 Construction of $R^i F$

The first and foremost question is the existence of the i-th derived functor. It turns out that this is closely related to the condition $(*)$ in the statement of Theorem 2.1 above that needs to be specified.

Definition 2.4 Let $I \in \mathcal{A}$ be an object of an abelian category. We call I *injective* if the functor $\mathrm{Hom}(\cdot, I)$ is exact (note that in general $\mathrm{Hom}(\cdot, I)$ is only left exact). Moreover, we say that *$\mathcal{A}$ has enough injectives* if for any object $A \in \mathcal{A}$, there exists an injective object I such that $0 \longrightarrow A \longrightarrow I$ is exact.

Example 2.2 (1) When $\mathcal{A} = \mathcal{G}$, the category of abelian groups, I is injective if and only if I is divisible, for instance, $\mathbb{Q}$ or $\mathbb{Q}/\mathbb{Z}$ (because the quotient of a divisible group is divisible). (2) $\mathrm{Sh}(X, \mathcal{G})$ and $\mathrm{Sh}(\mathbf{k}_X)$ have enough injectives (see Corollary 2.3 in Section 2 in Chapter III in [28]).

The following exercise is a standard fact in homological algebra, and it is also the first step is constructing derived functors.

Exercise 2.3 Suppose that $\mathcal{A}$ has enough injectives. For any object $A \in \mathcal{A}$, there exists a long exact sequence with I_i injective,

$$0 \longrightarrow A \longrightarrow I_0 \xrightarrow{f} I_1 \xrightarrow{f^{(1)}} \cdots, \tag{2.1}$$

where this exact sequence is called an *injective resolution of A in $\mathcal{A}$.*

Now let us construct the functor $R^i F$ that acts on any object $A \in \mathcal{A}$ when $\mathcal{A}$ has enough injectives. First, truncate term A from *any* injective resolution (2.1) of A, that is, one gets the following complex (not necessarily exact!):

$$0 \longrightarrow I_0 \xrightarrow{f} I_1 \xrightarrow{f^{(1)}} \cdots$$

Note that we won't lose any information because $\ker(f) = A$. Second, apply the functor F to this truncated sequence and get

$$0 \longrightarrow F(I_0) \xrightarrow{F(f)} F(I_1) \xrightarrow{F(f^{(1)})} \cdots$$

Note that this is still a complex (because F is assumed to be additive) but, again, not necessarily exact. Label the starting $F(I_0)$ as the degree-0 term, then **define**

$$R^i F(A) := \frac{\ker(F(f^{(i)}))}{\mathrm{im}(F(f^{(i-1)}))} = h^i(F(I_\bullet)). \tag{2.2}$$

Remark 2.2 (1) It is routine to check that the resulting $R^i F(A)$ is independent of the choice of injective resolutions of A. (2) It is easy to see that $R^0 F = F$, so it satisfies (1) in Theorem 2.1. In fact, since F is left exact, it preserves kernels. Since $\ker(f) = A$, one obtains that $F(A) = h^0(F(I_\bullet)) = R^0 F(A)$.

2.1.2 Computation of $R^i F$

Although we have constructed the derived functors $R^i F$, using injective resolutions to carry out concrete computations is almost impossible in practice. Here is a nicer class of objects that can be used for computations.

Definition 2.5 Let $F : \mathcal{A} \longrightarrow \mathcal{B}$ be a left exact functor. An object $X \in \mathcal{A}$ is called *F-acyclic* if $R^i F(X) = 0$ for all $i \geq 1$.

Example 2.3 If X is injective, then X is F-acyclic for *any* left exact functor F. In fact, we have a simple injective resolution of X,

$$0 \longrightarrow X \longrightarrow X \longrightarrow 0 \longrightarrow 0 \ldots$$

which implies that $R^i F(X) = 0$ for all $i \geq 1$ from the complex $0 \longrightarrow F(X) \longrightarrow 0 \longrightarrow \cdots$.

In the case of sheaves, we have the following definition.

Definition 2.6 An object $\mathcal{F} \in \mathrm{Sh}(X, \mathcal{G})$ is *flabby* if for every open subset $U \subset X$, the restriction map $F(X) \longrightarrow F(U)$ is surjective. It follows that, for any open subset $V \subset X$ such that $U \subset V$, the restriction map $F(V) \longrightarrow F(U)$ is surjective since restriction maps commute: $\mathrm{res}_{V,U} \circ \mathrm{res}_{X,V} = \mathrm{res}_{X,U}$.[4]

Exercise 2.4 Flabby sheaf is $\Gamma(X, \cdot)$-acyclic and also f_*-acyclic.

Then the following lemma (see Proposition 1.2 A. in Section 1 in Chapter III in [28]) shows that using F-acyclic resolutions we can also compute $R^i F$.

Lemma 2.1 *Let* $0 \longrightarrow A \longrightarrow X_1 \longrightarrow X_2 \longrightarrow \cdots$ *be an F-acyclic resolution, i.e., the sequence is exact and X_i is F-acyclic for any i. Then $R^i F(A) \simeq h^i(F(X_\bullet))$.*

Proposition 2.1 *Let* $f : X \longrightarrow Y$ *be a continuous map between topological spaces X and Y, and $\mathcal{F} \in \mathrm{Sh}(X, \mathcal{G})$. Then $R^i f_*(\mathcal{F})$ is a sheaf associated to the presheaf* $U \longrightarrow h^i(f^{-1}(U); \mathcal{F})(= R^i \Gamma(f^{-1}(U))(\mathcal{F}))$.

Example 2.4 We know that $\Gamma(X, \cdot)$ can be regarded as f_* for $f : X \longrightarrow Y$ where $Y = \{pt\}$. Then Proposition 2.1 can be equivalently recast as

$$R^i \Gamma(X, \cdot) = h^i(X, \cdot).$$

Apply this to $\mathcal{F} = \mathbf{k}_X$; then the right-hand side is the i-th cohomology group of the space X, while the left-hand side is an algebraic object coming from a bigger machinery, which constructs the i-th derived functor explained earlier. So we should think of these *higher degree* derived functors as being algebraic analogs of the *higher degree* cohomology groups. Moreover, if $Y = \mathbb{R}$ and $\mathcal{F} = \mathbf{k}_X$, then Proposition 2.1 recovers the classical Morse theory when it is applied to the open intervals $U = (-\infty, \lambda)$ for $\lambda \in \mathbb{R}$.

2.2 Derived Category

In this section, we will give the definition of a derived category. Let us start from specifying some subcategories of $\mathrm{Com}(\mathcal{A})$.

---

[4]There is another notion, called *soft*, which is defined as follows: for any closed subset $Z \subset X$, the restriction map $\mathcal{F}(X)$ every $F(Z)$ is surjective. In particular, when X is locally compact, every flabby sheaf is soft. Moreover, soft sheaf is $f_!$-acyclic.

(1) $\text{Com}^+(\mathcal{A}) = \{X_\bullet \in \text{Com}(\mathcal{A}) \mid X_i = 0 \text{ for } i \ll 0\};$
(2) $\text{Com}^-(\mathcal{A}) = \{X_\bullet \in \text{Com}(\mathcal{A}) \mid X_i = 0 \text{ for } i \gg 0\};$
(3) $\text{Com}^b(\mathcal{A}) = \{X_\bullet \in \text{Com}(\mathcal{A}) \mid X_i = 0 \text{ for } |i| \gg 0\};$
(4) $\text{Com}^{[m,n]}(\mathcal{A}) = \{X_\bullet \in \text{Com}(\mathcal{A}) \mid X_i = 0 \text{ for } i \notin [m, n]\}.$

Observe that any chain map (or morphism) $f : X_\bullet \longrightarrow Y_\bullet$ induces a map $h^i(f) :$ $h^i(X_\bullet) \longrightarrow h^i(Y_\bullet)$. Recall that f is called a *quasi-isomorphism* if $h^i(f)$ is an isomorphism for each $i \in \mathbb{Z}$. We call $X_\bullet$ and $Y_\bullet$ *quasi-isomorphic* if there exists a quasi-isomorphism between $X_\bullet$ and $Y_\bullet$.

Remark 2.3 For $X_\bullet$ and $Y_\bullet$ to be quasi-isomorphic, it is important that the isomorphism between $h^i(X_\bullet)$ and $h^i(Y_\bullet)$ is induced by a morphism $f : X_\bullet \longrightarrow Y_\bullet$. Only requiring that $h^i(X_\bullet) \simeq h^i(Y_\bullet)$ does *not* guarantee the existence of a quasi-isomorphism between $X_\bullet$ and $Y_\bullet$.

The idea of constructing the derived category $\mathcal{D}(\mathcal{A})$ is to formally invert all the quasi-isomorphisms. An analogous situation is encountered in from commutative algebra. Let R be a commutative ring and $S \subset R$ be a multiplicative subset. We can form the new ring (R, S) (or R_S), known as the *localization by* S. Each element in (R, S) is an equivalence class of a symbol $\frac{r}{s}$. Certainly, we have a canonical map $f : R \longrightarrow (R, S)$. More importantly, (R, S) has the following universality property: given any $g : R \longrightarrow R'$ that maps r to an invertible element $g(r)$ in R', there exists a unique map $h : (R, S) \longrightarrow R'$ such that $g = h \circ f$, that is, g factors through (R, S).

Definition 2.7 (or Theorem) For an abelian category $\mathcal{A}$, there exist a category denoted by $\mathcal{D}(\mathcal{A})$ and a functor $F : \text{Com}(\mathcal{A}) \longrightarrow \mathcal{D}(\mathcal{A})$, such that for any functor $G : \text{Com}(\mathcal{A}) \longrightarrow C$, where G maps any quasi-isomorphism in $\text{Com}(\mathcal{A})$ to an isomorphism in C, there exists a unique (up to isomorphisms) functor $H : \mathcal{D}(\mathcal{A}) \longrightarrow C$ such that the following diagram commutes

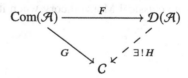

This category $\mathcal{D}(\mathcal{A})$ is called the *derived category of* $\mathcal{A}$.

More explicitly, $\text{Obj}(\mathcal{D}(\mathcal{A})) = \text{Obj}(\text{Com}(\mathcal{A}))$. For any $X, Y \in \text{Obj}(\mathcal{D}(\mathcal{A}))$, a morphism in $\text{Mor}_{\mathcal{D}(\mathcal{A})}(X, Y)$ is an equivalence class represented by a *roof* of the form

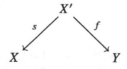

where s is a quasi-isomorphism (hence invertible in $\mathcal{D}(\mathcal{A})$) and f is a morphism in $\mathcal{A}$. The equivalence relation can be found in Definition 1.6.2 in [32].

Example 2.5 For each $i \in \mathbb{Z}$, consider functor $h^i : \mathrm{Com}(\mathcal{A}) \longrightarrow \mathcal{A}$. By definition, h^i maps each quasi-isomorphism to an isomorphism (in $\mathcal{A}$). By Definition/Theorem 2.7, there exists a unique functor (still denoted by) $h^i : \mathcal{D}(\mathcal{A}) \longrightarrow \mathcal{A}$, that is, the cohomology functor is still well defined on $\mathcal{D}(\mathcal{A})$.

There are variants based on Definition/Theorem 2.7 in which one starts with different subcategories of $\mathrm{Com}(\mathcal{A})$. For instance, starting with $\mathrm{Com}^+(\mathcal{A})$, we can define $\mathcal{D}^+(\mathcal{A})$, similarly to $\mathcal{D}^-(\mathcal{A})$, $\mathcal{D}^b(\mathcal{A})$ and $\mathcal{D}^{[m,n]}(\mathcal{A})$. The following theorem (see Proposition 2.30 in [30]) provides easy descriptions of these variants.

Theorem 2.2 *There exists an equivalence of categories*

$$\mathcal{D}^+(\mathcal{A}) \simeq \left\{ X_\bullet \in \mathcal{D}(\mathcal{A}) \mid h^i(X_\bullet) = 0 \text{ for } i \ll 0 \right\}$$

where the right-hand side is a full subcategory of $\mathcal{D}(\mathcal{A})$. Similar conclusions hold for $\mathcal{D}^-(\mathcal{A})$, $\mathcal{D}^b(\mathcal{A})$, and $\mathcal{D}^{[m,n]}(\mathcal{A})$.

In particular, since $\mathcal{A} \simeq \mathrm{Com}^{[0,0]}(\mathcal{A})$ by just identifying X with the complex

$$X_\bullet = \cdots \longrightarrow 0 \longrightarrow X \longrightarrow 0 \longrightarrow \cdots$$

where X is concentrated at degree 0, Theorem 2.2 says that, up to an equivalence, $\mathcal{D}^{[0,0]}(\mathcal{A})$ consists of all complexes $Y_\bullet \in \mathcal{D}(\mathcal{A})$ such that $h^i(Y_\bullet) = 0$ for all $|i| \geq 1$. In particular, $X_\bullet$ above provides such an example. In other words, one has an embedding $\mathcal{A} \hookrightarrow \mathcal{D}^{[0,0]}(\mathcal{A})$. In fact, this is an equivalence, see 2. Proposition in Section 5 in Chapter III in [21]. Thus we can regard any $\mathcal{A}$ as a segmental information in a bigger frame, its derived category.

2.3 Upgrade to Functor RF

In this section, we will give the definition of RF, called the derived functor of F[5] that was introduced in Remark 2.1. Let $\mathcal{A}$ be an abelian category and $\mathrm{Com}(\mathcal{A})$ be the associated chain complex category.

Definition 2.8 Let $f_\bullet, g_\bullet : X_\bullet \longrightarrow Y_\bullet$ be two morphisms in $\mathrm{Com}(\mathcal{A})$. We say $f_\bullet$ is *homotopic to $g_\bullet$* (and simply write $f \sim g$), if there exist maps $h_\bullet : X_\bullet \longrightarrow Y_{\bullet-1}$ such that in the following diagram

---

[5] Warning on notations: in the previous section, we have defined a family of functors $R^i F : \mathcal{A} \longrightarrow \mathcal{B}$ for $i \geq 0$ (see (2.2)), while RF here should be regarded as a new definition/notation.

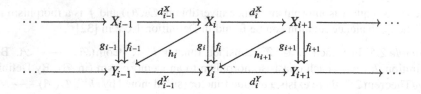

for any degree $i \in \mathbb{Z}$,

$$f_i - g_i = h_{i+1} \circ d_i^X + d_{i-1}^Y \circ h_i.$$

Moreover, two chain complexes $X_\bullet$ and $Y_\bullet$ are *homotopic* (denoted by $X_\bullet \sim Y_\bullet$) if there exist morphisms $f_\bullet : X_\bullet \longrightarrow Y_\bullet$ and $g_\bullet : Y_\bullet \longrightarrow X_\bullet$ such that $g \circ f \sim \mathbb{1}_{X_\bullet}$ and $f \circ g \sim \mathbb{1}_{Y_\bullet}$.

Exercise 2.5 If $f \sim g : X_\bullet \longrightarrow Y_\bullet$, then $h^i(f) = h^i(g) : h^i(X_\bullet) \longrightarrow h^i(Y_\bullet)$.

Exercise 2.6 Let $0 \longrightarrow A \longrightarrow I_\bullet$ and $0 \longrightarrow A \longrightarrow J_\bullet$ be two injective resolutions. Then $I_\bullet \sim J_\bullet$.

Exercise 2.7 If $X_\bullet \sim Y_\bullet$, then $[X_\bullet] \simeq [Y_\bullet]$ in $\mathcal{D}(\mathcal{A})$ where $[-]$ denotes the quasi-isomorphic class in the derived category.

The following theorem can be regarded as the definition of the functor RF with domain $\mathcal{A}$ (and the concrete definition of RF is given in the proof). Later, in Theorem 2.4, the domain of RF can be upgraded to be $\mathcal{D}^+(\mathcal{A})$.

Theorem 2.3 (Relation between RF and R^iF) *Let $F : \mathcal{A} \longrightarrow \mathcal{B}$ be a left exact functor and assume that $\mathcal{A}$ has enough injectives. There exists a natural functor $RF : \mathcal{A} \longrightarrow \mathcal{D}^+(\mathcal{B})$ such that $h^i \circ RF \simeq R^iF$.*

Proof For any $A \in \mathcal{A}$, since $\mathcal{A}$ has enough injectives, we can find an injective resolution of $\mathcal{A}$ as follows,

$$0 \longrightarrow A \longrightarrow I_0 \longrightarrow I_1 \longrightarrow \cdots$$

Denote by $I_\bullet$ the chain complex $\cdots \longrightarrow 0 \longrightarrow I_0 \longrightarrow I_1 \longrightarrow \cdots$, and **define**

$$RF(A) := [F(I_\bullet)] \in \mathcal{D}^+(\mathcal{B}). \tag{2.3}$$

We need to show that the definition (2.3) is correct. So, let $J_\bullet$ be another injective resolution. By Exercise 2.6, $I_\bullet \sim J_\bullet$. Then $F(I_\bullet) \sim F(J_\bullet)$. By Exercise 2.7, $[F(I_\bullet)] \simeq [F(J_\bullet)]$. Finally since h^i is well defined in $\mathcal{D}^+(\mathcal{B})$, by definition (2.2) of R^iF, the conclusion follows. $\square$

In fact, we can extend the domain of Theorem 2.3 from $\mathcal{A}$ to $\mathcal{D}^+(\mathcal{A})$ since $\mathcal{A} \simeq \mathcal{D}^{[0,0]}(\mathcal{A}) \hookrightarrow \mathcal{D}^+(\mathcal{A})$.

Theorem 2.4 *Let $F : \mathcal{A} \longrightarrow \mathcal{B}$ be a left exact functor and assume that $\mathcal{A}$ has enough injectives. Then there exists a natural functor $RF : \mathcal{D}^+(\mathcal{A}) \longrightarrow \mathcal{D}^+(\mathcal{B})$ such that it is an extension of RF on $\mathcal{A}$. This upgraded functor RF is called the derived functor of F.*

Proof The proof essentially provides an algorithm to construct a complex $I_\bullet$ for a given $Z_\bullet \in \text{Com}^+(\mathcal{A})$ such that $Z_\bullet \hookrightarrow I_\bullet$, where the embedding is degree-wise and each I_i is injective. Moreover, such $I_\bullet$ is unique up to homotopy. Then we will define

$$RF(Z_\bullet) = [F(I_\bullet)].$$

Explicitly, we can inductively find the following diagram by taking advantage of the fact that $\mathcal{A}$ has enough injectives.

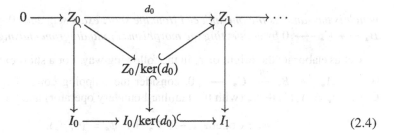

$$(2.4)$$

where I_1 can be explicitly written as $I_1 = (Z_1 \oplus I_0/\ker(d_0))/(Z_0/\ker(d_0))$. $\square$

Remark 2.4 (a) To understand (2.4) better, readers are encouraged to consider the simplest case of the complex $Z_\bullet = (0 \longrightarrow A \longrightarrow 0)$ i.e., there is only one term $A \in \mathcal{A}$. Then the diagram (2.4) gives

$$X_\bullet = I_0 \longrightarrow I_0/A \longrightarrow I_1 \longrightarrow \dots \qquad (2.5)$$

where I_i is injective. An interesting observation is that $(0 \longrightarrow A \longrightarrow 0)$ and $I_\bullet$ in (2.5) are quasi-isomorphic, but *not* homotopic! In fact, any morphism $g_\bullet : I_\bullet \longrightarrow (0 \longrightarrow A \longrightarrow 0)$ satisfying the homotopy relation will result in an isomorphism $A \simeq I_0$ (because any h_i required in Definition 2.8 is 0). (b) Definition 8.12 in [61] provides an alternative approach to define the derived functor $RF : \mathcal{D}^b(\mathcal{A}) \longrightarrow \mathcal{D}^b(\mathcal{B})$ whenever $\mathcal{A}$ has enough injectives. Explicitly, for an element $Z_\bullet \in \text{Com}^b(\mathcal{A})$, one can associate to it a Cartan-Eilenberg resolution (see Definition 8.4 in [61]). As a double complex, this resolution is formed by injective objects. Proposition 8.7 in [61] proves that the total complex of this Cartan-Eilenberg resolution is actually quasi-isomorphic to $Z_\bullet$. Then RF is defined by associating to $Z_\bullet$ the image under F of the total complex of a Cartan-Eilenberg resolution of $Z_\bullet$.

2.4   Triangulated Structure

The derived category $\mathcal{D}(\mathcal{A})$ has a richer structures than the chain complex category $\text{Com}(\mathcal{A})$. One extra structure is called the *triangulated structure*. A starting observation is that in the category $\text{Com}(\mathcal{A})$, a short exact sequence $0 \longrightarrow A_\bullet \longrightarrow B_\bullet \longrightarrow C_\bullet \longrightarrow 0$ does not necessarily admit any map from $C_\bullet$ (back) to $A_\bullet$. In sharp contrast, once we pass to the derived category $\mathcal{D}(\mathcal{A})$, we have the following theorem.

Theorem 2.5 *In $\mathcal{D}(\mathcal{A})$, there exists a well-defined map $\tau_\bullet : C_\bullet \longrightarrow A_\bullet[1]$ such that after applying the cohomological functor h^i to $A_\bullet \longrightarrow B_\bullet \longrightarrow C_\bullet \xrightarrow{\tau_\bullet} A_\bullet[1]$, one gets a long exact sequence*

$$\cdots \longrightarrow h^i(A_\bullet) \longrightarrow h^i(B_\bullet) \longrightarrow h^i(C_\bullet) \xrightarrow{h^i(\tau_\bullet)} h^{i+1}(A_\bullet) \longrightarrow \cdots$$

which is the same as the one induced from the short exact sequence $0 \longrightarrow A_\bullet \longrightarrow B_\bullet \longrightarrow C_\bullet \longrightarrow 0$ by connecting the morphisms δ_i via diagram chasing.

Let us elaborate the origin of $\tau_\bullet$ in the following way. For a short exact sequence $0 \longrightarrow A_\bullet \xrightarrow{\alpha} B_\bullet \xrightarrow{\beta} C_\bullet \longrightarrow 0$, consider the mapping cone of α, defined by $\text{Cone}(\alpha)_\bullet = A_\bullet[1] \oplus B_\bullet$ (with the familiar boundary operator) and the map

$$\phi_\bullet : \text{Cone}(\alpha)_\bullet \longrightarrow C_\bullet \quad \text{by} \quad \phi_\bullet = (0, \beta_\bullet).$$

One can check (**Exercise**) that such $\phi_\bullet$ is a quasi-isomorphism (Proposition 1.7.5 in [32]). Then in $\mathcal{D}(\mathcal{A})$, we can safely replace $C_\bullet$ with $\text{Cone}(\alpha)_\bullet$. An immediate advantage is that there exists a well-defined map $\text{Cone}(\alpha)_\bullet \longrightarrow \mathcal{A}_\bullet[1]$: just take the projection onto $\mathcal{A}_\bullet[1]$. Therefore, in $\mathcal{D}(\mathcal{A})$ we have the sequence

$$A_\bullet \longrightarrow B_\bullet \longrightarrow C_\bullet \xrightarrow{\tau_\bullet} A_\bullet[1], \tag{2.6}$$

where the morphism $\tau_\bullet$ is defined by first identifying $C_\bullet$ with $\text{Cone}(\alpha)_\bullet$ and then projecting. Moreover, applying the cohomology functor, (2.6) gives a long exact sequence, and we can check that the induced $\tau_\bullet$ is the same as the morphism $\delta_\bullet$ that is constructed in homological algebra.

Definition 2.9 *A distinguished triangle* in $\mathcal{D}(\mathcal{A})$ is a sequence

$$X_\bullet \longrightarrow Y_\bullet \longrightarrow Z_\bullet \longrightarrow X_\bullet[1]$$

such that it is quasi-isomorphic (for each position) to some sequence (2.6) — the sequence/triangle derived from a short exact sequence.

Traditionally, whenever a category C is endowed with an automorphism $T : C \longrightarrow C$ (for instance, a degree shift in $\text{Com}(\mathcal{A})$) and a family of distinguished triangles which satisfy certain properties, we call C a *triangulated category*. The theory of triangulated categories is extensively developed (see Section 1.5 in Chapter 1 in [32]). In particular, every derived category is a triangulated category.

Proposition 2.2 *One has following properties of distinguished triangles.*

(0) *Every morphism $X \xrightarrow{f} Y$ can be completed to a distinguished triangle $X \xrightarrow{f} Y \longrightarrow Z \xrightarrow{+1}$.*

(1) *(Definition 1.5.2 and Proposition 1.5.6 in [32]) If $X_\bullet \longrightarrow Y_\bullet \longrightarrow Z_\bullet \longrightarrow X_\bullet[1]$ is a distinguished triangle as defined in Definition 2.9, then we have a long exact sequence*

$$\cdots \longrightarrow h^i(X_\bullet) \longrightarrow h^i(Y_\bullet) \longrightarrow h^i(Z_\bullet) \longrightarrow h^{i+1}(X_\bullet) \longrightarrow \cdots$$

(2) For any left exact functor $F : \mathcal{A} \longrightarrow \mathcal{B}$, RF maps a distinguished triangle into a distinguished triangle.

Remark 2.5 (Important!) A functor from $\mathcal{D}(\mathcal{A})$ to $\mathcal{D}(\mathcal{B})$ that satisfies (2) in Proposition 2.2 is called an *exact functor in the derived category*. Now, we arrive at the point that really clarifies the "exactness" of a derived functor mentioned in Remark 2.1. Recall that in the category $\mathcal{A}$, an exact functor (from $\mathcal{A}$ to $\mathcal{B}$) takes any short exact sequence into a short exact sequence. If not, then we get a long exact sequence by successively taking $R^i F$ to detect the non-exactness. However, by the discussion above, a distinguished triangle is an analog of a short exact sequence in $\mathcal{D}(\mathcal{A})$. Upgrade F to its derived functor RF. Then RF preserves distinguished triangles, which, by definition, means that it is (always) exact in a derived category.

2.5 Applications to Sheaves

In this section, we will apply the abstract constructions developed in the previous sections to a concrete category, the category of sheaves of groups.

2.5.1 Base Change Formula

In Example 2.1, given a continuous map $f : X \longrightarrow Y$, we have defined sheaf operators f_*, f^{-1} and $f_!$ (for this the spaces X and Y are required to be locally compact). Let us describe the stalk of $f_!$ first. It behaves better than the stalk of f_* in general.

Proposition 2.3 (Proposition 2.5.2 in [32]) *Let $f : X \longrightarrow Y$ be a continuous map function where X and Y are locally compact spaces. For any $y \in Y$ and $\mathcal{F} \in \mathrm{Sh}(X)$, there exists a canonical isomorphism*

$$(f_!\mathcal{F})_y \simeq \Gamma_c(f^{-1}(y), \mathcal{F}|_{f^{-1}(y)}).$$

The functors $\Gamma(X, \cdot)$, $\Gamma_c(X, \cdot)$, f_* and $f_!$ are all left exact. Therefore, we have their derived functors, denoted by $R\Gamma(X, \cdot)$, $R\Gamma_c(X, \cdot)$, Rf_* and $Rf_!$, respectively, which are well defined in the corresponding derived categories. Note that f^{-1}, by definition, is already exact, so $Rf^{-1} = f^{-1}$. Moreover, there is a useful fact called the *Grothendieck composition formula* which addresses the issue of the composition of two functors $G \circ F$ where $\mathcal{A} \xrightarrow{F} \mathcal{B} \xrightarrow{G} C$. Under certain conditions (for instance, that F maps F-acyclic objects into G-acyclic objects), we get

$$R(G \circ F) = RG \circ RF. \tag{2.7}$$

Example 2.6 Let $f : X \longrightarrow Y$ be a continuous map between two topological spaces. Consider the composition

$$\mathrm{Sh}(X) \xrightarrow{f_*} \mathrm{Sh}(Y) \xrightarrow{\Gamma(Y, \cdot)} \mathcal{A}b.$$

By Exercise 1.16 (d) in [28], f_* maps any flabby sheaf into a flabby sheaf. Moreover, using Exercise 2.4 above asserting that flabby sheaves are $\Gamma(X, \cdot)$-acyclic, we apply the composition formula and get

$$R\Gamma(Y, Rf_*\mathcal{F}) = R(\Gamma(Y, f_*\mathcal{F})) = R\Gamma(X, \mathcal{F}).$$

The main theorem in this subsection is called *base change* formula.

Proposition 2.4 *Suppose that we have a **cartesian** square*

$$
\begin{array}{ccc}
X & \xleftarrow{\;\beta\;} & X' \\
{\scriptstyle f}\downarrow & & \downarrow{\scriptstyle g} \\
Y & \xleftarrow{\;\alpha\;} & Y'
\end{array}
$$

that is, $X' \simeq X \times_Y Y'$. Then there exists a canonical isomorphism such that, for any $\mathcal{F} \in \mathcal{D}^+(\mathrm{Sh}(X))$, $\alpha^{-1}(Rf_!\mathcal{F}) \simeq Rg_!(\beta^{-1}\mathcal{F})$.

Remark 2.6 Since $R\alpha^{-1} = \alpha^{-1}$ and $R\beta^{-1} = \beta^{-1}$, if we can prove that

$$\alpha^{-1}(f_!\mathcal{F}) \simeq g_!(\beta^{-1}\mathcal{F}), \tag{2.8}$$

then since β^{-1} sends flabby sheaves to $g_!$-acyclic ones, the composition formula (2.7) yields

$$\alpha^{-1}(Rf_!\mathcal{F}) = R(\alpha^{-1})(Rf_!(\mathcal{F})) = R(\alpha^{-1} \circ f_!)(\mathcal{F})$$

$$\simeq R(g_! \circ \beta^{-1})(\mathcal{F}) = Rg_!(R\beta^{-1}(\mathcal{F})) = Rg_!(\beta^{-1}\mathcal{F}),$$

which is our desired conclusion in Proposition 2.4.

Proof Our plan to prove Proposition 2.4 is by first assuming that the following assertion (whose proof needs a concept call *adjoint functors* which will be introduced later) is valid.

There exists a canonical morphism $\alpha^{-1}(f_!\mathcal{F}) \longrightarrow g_!(\beta^{-1}\mathcal{F})$. $\qquad$ (2.9)

Then by Remark 2.6, one only needs to check (2.8) at the level of stalks. In fact, for any point $y' \in Y'$, we have

$$((\alpha^{-1} f_!)(\mathcal{F}))_{y'} = (f_!\mathcal{F})_{\alpha(y')} = \Gamma_c(f^{-1}(\alpha(y')), \mathcal{F}). \qquad (2.10)$$

On the other hand,

$$((g_!\beta^{-1})(\mathcal{F}))_{y'} = \Gamma_c(g^{-1}(y'), \beta^{-1}(\mathcal{F})). \qquad (2.11)$$

Now, (2.10) is the same as (2.11) due to the assumption that the square is cartesian, and β provides an isomorphism between the fibers $g^{-1}(y')$ and $f^{-1}(\alpha(y'))$. $\qquad \square$

2.5.2 Adjoint Relation

Given two functors $F : \mathrm{Sh}(X) \longrightarrow \mathrm{Sh}(Y)$ and $G : \mathrm{Sh}(Y) \longrightarrow \mathrm{Sh}(X)$, we say F and G form an *adjoint pair* if for any $\mathcal{F} \in \mathrm{Sh}(X)$ and $\mathcal{G} \in \mathrm{Sh}(Y)$,

$$\mathrm{Hom}_{\mathrm{Sh}(Y)}(F(\mathcal{F}), \mathcal{G}) = \mathrm{Hom}_{\mathrm{Sh}(X)}(\mathcal{F}, G(\mathcal{G})). \qquad (2.12)$$

More precisely, F is the[6] left adjoint functor of G, and G is the right adjoint functor of F. One of the most important properties of adjoint functors is that they provide canonical morphisms. Let us elaborate on this by the following lemma.

Lemma 2.2 (Exercise I.2(i) **in [32])** *For any* $\mathcal{F} \in \mathrm{Sh}(X)$ *and* $\mathcal{G} \in \mathrm{Sh}(Y)$, *if* F *and* G *are adjoint functors as in (2.12), then we have morphisms*

$$\mathcal{F} \longrightarrow (G \circ F)(\mathcal{F}) \quad and \quad (F \circ G)(\mathcal{G}) \longrightarrow \mathcal{G}.$$

Note that the orders are important!

Example 2.7 The functors f_* and f^{-1} are adjoint functors. Explicitly, we have

$$\mathrm{Hom}(f^{-1}\mathcal{F}, \mathcal{G}) = \mathrm{Hom}(\mathcal{F}, f_*\mathcal{G}). \qquad (2.13)$$

Now let us go back to the gap in the proof of Proposition 2.4.

---

[6]It is possible that there exist more than one left adjoint of G. But Exercise I.2 (ii) in [32] says that all of them are isomorphic, so the left adjoint will be unique (up to isomorphisms). The same is true for the right adjoint.

Proof (*of the assertion* (2.9)) We will first show that, for any $\mathcal{F} \in \mathrm{Sh}(X')$, there exists a morphism

$$(f_! \circ \beta_*)(\mathcal{F}) \longrightarrow (\alpha_* \circ g_!)(\mathcal{F}). \tag{2.14}$$

In fact, for any open subset $U \subset Y$, a section s of $(f_! \circ \beta_*)(\mathcal{F})(U)$ satisfies

$$\mathrm{supp}(s) \subset \beta^{-1}(V)$$

where $V \subset f^{-1}(U)$ and $f : V \longrightarrow U$ is proper. Then we have the following picture

$$
\begin{array}{ccc}
V & \overset{\beta}{\longleftarrow} & \beta^{-1}(V) \\
{\scriptstyle\text{proper}}\big\downarrow & & \big\downarrow{\scriptstyle\text{proper}} \\
U & \overset{\alpha}{\longleftarrow} & \alpha^{-1}(U)
\end{array}
$$

where the properness on the right vertical arrow comes from our hypothesis that the square is cartesian (so g provides an isomorphism on fibers). Therefore, s trivially defines a section on $(\alpha_* \circ g_!)(\mathcal{F})(U)$, by definition.

Next, we will play with the adjoint relations. Note that for any $\mathcal{G} \in \mathrm{Sh}(X)$, by Lemma 2.2 and (2.13),

$$f_!(\mathcal{G}) \longrightarrow (f_! \circ \beta_* \circ \beta^{-1})(\mathcal{G}) \longrightarrow (\alpha_* \circ g_! \circ \beta^{-1})(\mathcal{G}) \tag{2.15}$$

where the last arrow comes from (2.14). Therefore, by the adjoint relationship,

$$\mathrm{Hom}(\alpha^{-1}(f_!(\mathcal{G})), g_!(\beta^{-1}(\mathcal{G}))) = \mathrm{Hom}(f_!(\mathcal{G}), (\alpha_* \circ g_! \circ \beta^{-1})(\mathcal{G})).$$

By (2.15), we know that there exists at least one non-trivial morphism on the right-hand side. Then it corresponds to a morphism on the left-hand side, and this provides the sought-for morphism in the lemma.   □

Since Hom is well-known to be left exact, its derived functor $R\mathrm{Hom}$ is well-defined in the corresponding derived category. Moreover, in this derived category, we also have an adjoint relation, for instance,

$$R\mathrm{Hom}(f^{-1}\mathcal{F}, \mathcal{G}) = R\mathrm{Hom}(\mathcal{F}, Rf_*\mathcal{G}). \tag{2.16}$$

By (1) in Theorem 2.1, equation (2.13) can be regarded as the degree-0 derived version of equation (2.16).

Remark 2.7 An interesting phenomenon is that, in general, the functor $f_!$ does not admit any adjoint functor on the level of the category of sheaves, except for the elementary case where f is an open/closed embedding. In order to define an adjoint functor (more precisely, the right adjoint), we need to pass to its derived functor $Rf_!$. This story involves Verdier duality, and it will not be elaborated in detail in this book. The interested readers may consult Chapter III in [32].

2.6 Persistence k-Modules

The theory of persistence **k**-modules provides a translation from topological/geometrical questions to combinatorial questions via a novel algebraic structure. It started with the groundbreaking work by Zomorodian and Carlsson in 2005 [63], with a motivation in a subject nowadays called topological data analysis. This subject provides a new method to study shapes, and it allows modern computers as well as powerful computational programs to enter this field and help people "view" geometry from topology.

Definition 2.10 A persistence **k**-module (V, π) consists of the following data:

- $V = \{V_t\}_{t \in \mathbb{R}}$ such that for each $t \in \mathbb{R}$, V_t is finite-dimensional over **k**;
- $\pi_{st} : V_s \longrightarrow V_t$ for any $s \leq t$ such that (i) $\pi_{tt} =$ identity and (ii) for any $s \leq t \leq r$, $\pi_{sr} = \pi_{tr} \circ \pi_{st}$.

Moreover, for practical applications, we will also assume following conditions:

- $V_s = 0$ for $s << 0$;
- (regularity) for all but finitely many points $t \in \mathbb{R}$, there exists a neighborhood U of t such that $\pi_{sr} : V_s \longrightarrow V_r$ is an isomorphism for any $s \leq r$ in U;
- (semi-continuity) for any $t \in \mathbb{R}$, π_{st} is an isomorphism for any s that is sufficiently close to t from *left*.

In the regularity condition above, denote by $\mathrm{Spec}(V, \pi)$ the collection of the jump points for which the neighborhood isomorphism condition is not satisfied.

Exercise 2.8 For $s >> 0$, $V_s = V_\infty$.

Example 2.8 A standard example of a persistent **k**-module is the so-called *interval-type* **k**-module. Fix an interval $(a, b]$

$$\mathbb{I}_{(a,b]} = \begin{cases} \mathbf{k}, & \text{if } t \in (a, b], \\ 0, & \text{otherwise,} \end{cases}$$

and π_{st} is the identity map on **k** whenever both $s, t \in (a, b]$, and 0 otherwise.

Example 2.9 One can obtain a persistence **k**-module from the classical Morse theory. Let X be a closed manifold and $f : X \longrightarrow \mathbb{R}$ be a Morse function. Define

$$V_t = H_*(\{f < t\}; \mathbf{k}).$$

Then, if $s < t$, the inclusion map $\{f < s\} \subset \{f < t\}$ induces a map on (filtered) homologies $\pi_{st} : V_s \longrightarrow V_t$. Moreover, $\mathrm{Spec}(V, \pi) = \{$critical values of $f\}$.

Example 2.10 Let (X, d) be a finite metric space. For any $t > 0$, set $R_t(X, d)$ to be a simplical complex (called the *Rips complex*) such that (i) $\{x_i\}$ are vertices and (ii) $\sigma \subset X$ is a simplex in $R_t(X, d)$ if $\mathrm{diam}(\sigma) < t$. Then $H_*(R_t(X, d); \mathbf{k})$ is a persistence $\mathbf{k}$-module.

Definition 2.11 Let (V, π) and (W, θ) be two persistence $\mathbf{k}$-modules. A (persistence) morphism between (V, π) and (W, θ) is an $\mathbb{R}$-family of $\mathbf{k}$-linear maps $A_t : V_t \longrightarrow W_t$ such that the following diagram commutes:

$$
\begin{array}{ccc}
V_s & \xrightarrow{\ \pi_{st}\ } & V_t \\
\downarrow{\scriptstyle A_s} & & \downarrow{\scriptstyle A_t} \\
W_s & \xrightarrow[\ \theta_{st}\ ]{} & W_t
\end{array}
$$

Example 2.11 There is *no* morphism from $\mathbb{I}_{(1,2]}$ to $\mathbb{I}_{(1,3]}$. But there exist non-trivial morphisms from $\mathbb{I}_{(2,3]}$ to $\mathbb{I}_{(1,3]}$.

Definition 2.12 Let (W, θ) be a persistence $\mathbf{k}$-module. $(V, \pi) \subset (W, \theta)$ is a persistence submodule if the inclusion is a morphism.

Example 2.12 $\mathbb{I}_{(2,3]} \subset \mathbb{I}_{(1,3]}$ is a persistence submodule. However, it does not have a direct complement (in the persistence sense).

Exercise 2.9 For any $a < b < c$, there exists an short exact sequence

$$
0 \longrightarrow \mathbb{I}_{(b,c]} \longrightarrow \mathbb{I}_{(a,c]} \longrightarrow \mathbb{I}_{(a,b]} \longrightarrow 0.
$$

Remark 2.8 Example 2.12 can be interpreted as asserting that $\mathbb{I}_{(a,c]}$ is *indecomposable*. In other words, the short exact sequence from Exercise 2.9 does not split. An analogous result in the world of sheaves is that every sheaf $\mathbf{k}_{(a,c]}$ is indecomposable. Let us give the definition of $\mathbf{k}_{(a,c]}$. For completeness, for any sheaf $\mathcal{F} \in \mathrm{Sh}(X)$, let us recall the notation $\mathcal{F}_*$ where $*$ is an open, a closed, or a locally closed subset of X. Note that a subset is called locally closed if it can be written as the intersection of an open subset and a closed subset. Let $A \subset X$ be a closed subset. Then $\mathcal{F}_A := j_* j^{-1} \mathcal{F}$, where $j : A \longrightarrow X$ is the inclusion. Let $U \subset X$ be an open subset. Then $\mathcal{F}_U := \ker(\mathcal{F} \longrightarrow \mathcal{F}_{X \setminus U})$. Let $Z \subset X$ be a locally closed subset: $Z = U \cap A$ for some open subset U of X and some closed subset A of X. Then $\mathcal{F}_Z := (\mathcal{F}_U)_A$. For the case of $\mathbf{k}_{(a,c]}$, apply the construction above to $\mathcal{F} = \mathbf{k}_{\mathbb{R}}$, the constant sheaf over $\mathbb{R}$, and the interval $Z = (a, c]$, which is equal to $(a, \infty) \cap (-\infty, c]$.

The interval-type $\mathbf{k}$-modules from Example 2.8 are more special and play a fundamental role, as explained by to the following structure theorem.

Theorem 2.6 (Normal Form Theorem) *For any persistence* **k**-*module* (V, π), *there exists a unique collection of intervals* $\mathcal{B} = \{(I_j = (a_j, b_j], m_j)\}$, *where* m_j *is the multiplicity of* $(a_j, b_j]$, *such that*

$$V = \bigoplus \mathbb{I}_{(a_j, b_j]}^{m_j}.$$

Definition 2.13 We call the collection of $\mathcal{B} = \mathcal{B}(V)$ from Theorem 2.6 the *barcode of* (V, π). Therefore, we have a well-defined association $V \mapsto \mathcal{B}(V)$.

There are various approaches to prove Normal Form Theorem (even for persistence **k**-modules with more general definitions). Interested readers can consult [12, 63]. One way to explain this structure theorem is via quiver representations. Recall that a quiver representation is planar directed graph where each vertex represents a finite-dimensional vector space (over **k**) and each arrow represents a linear map from its source to its target. Observe that a persistence **k**-module as defined in this section can be viewed as a quiver representation, where all the vertices lie on a line and arrows go successively in one direction. In fact, this is a special case of an A_n-type quiver. A classical theorem of Gabriel (see [20]) classifies quiver representations. Explicitly, a quiver representation can be decomposed into a direct sum of *indecomposable representations*. It turns out that, for A_n-type quivers, the indecomposable representations are the interval-type quivers. This again emphasizes the importance of the interval type **k**-modules. Moreover, the barcode is uniquely determined by the quiver representation.

2.7 Persistence Interleaving Distance

Before defining the interleaving distance, we need to talk about shift functors. Let (V, π) be a persistence **k**-module and $\delta > 0$. Denote by $(V[\delta], \pi[\delta])$ a new persistence **k**-module, where for each $t \in \mathbb{R}$,

$$V[\delta]_t = V_{t+\delta} \quad \text{and} \quad \pi[\delta]_{s,t} = \pi_{s+\delta, t+\delta}.$$

Moreover, due to the positivity of δ, we have a natural map

$$\Phi_V^\delta : (V, \pi) \longrightarrow (V[\delta], \pi[\delta]),$$

where for any $t \in \mathbb{R}$, $\Phi_V^\delta(t) = \pi_{t, t+\delta}$. For any morphism $F : V \longrightarrow W$, denote by $F[\delta] : V[\delta] \longrightarrow W[\delta]$ the δ-shifted morphism. Now, we are ready to define the interleaving distance, where the interleaving relation should be regarded as a shifted version of a persistence isomorphism.

Definition 2.14 Let (V, π) and (W, π') be two persistence **k**-modules and $\delta > 0$. We say that they are δ-*interleaved* if there exist morphisms $F : V \longrightarrow W[\delta]$ and $G : W \longrightarrow V[\delta]$ such that

$$G[\delta] \circ F = \Phi_V^{2\delta} \quad \text{and} \quad F[\delta] \circ G = \Phi_W^{2\delta}.$$

The interleaving distance between V and W is defined as

$$d_{\text{int}}(V, W) = \inf\{\delta > 0 \mid V \text{ and } W \text{ are } \delta\text{-interleaved}\}.$$

Exercise 2.10 Show that $d_{\text{int}}(V, W) < \infty$ if and only if $\dim(V_\infty) = \dim(W_\infty)$.

Exercise 2.11 Fix any $n \in \mathbb{N}$. Show that

$$(\{\text{persistence } \mathbf{k}\text{-modules with } \dim(V_\infty) = n\} \,/\text{isom.}, d_{\text{int}})$$

is a metric space.

Example 2.13 (How to get an interleaving relation?) Let $a < b$ and $c < d$. Consider the interval-type **k**-modules $\mathbb{I}_{(a,b]}$ and $\mathbb{I}_{(c,d]}$. There are two strategies to get interleaving relations:

(I) Let $\delta = \max\left(\frac{b-a}{2}, \frac{d-c}{2}\right)$. Note that then $b - 2\delta \leq a$, so the intervals $(a-2\delta, b-2\delta]$ and $(a, b]$ are disjoint. Therefore, $\Phi_{\mathbb{I}_{(a,b]}}^{2\delta} = 0$, which implies that we can choose 0-morphisms to form an interleaving relation. Denote this δ by δ_{I}.

(II) Let $\delta = \max(|a - c|, |b - d|)$. Then

$$a - 2\delta \leq c - \delta \leq a \quad \text{and} \quad b - 2\delta \leq d - \delta \leq b.$$

One gets a well-defined morphism from $\mathbb{I}_{(a,b]} \longrightarrow \mathbb{I}_{(c-\delta,d-\delta]} \longrightarrow \mathbb{I}_{(a-2\delta,b-2\delta]}$. Denote this δ by δ_{II}.

Hence, by definition, $d_{\text{int}}(\mathbb{I}_{(a,b]}, \mathbb{I}_{(c,d]}) \leq \min(\delta_{\text{I}}, \delta_{\text{II}})$. In fact, equality holds.

Example 2.14 (Interleaving relations from geometry) Let M be a closed manifold and $h \in C^\infty(M)$. Denote $\|h\| = \max_M |h|$ and $V(f) = H_*(\{f < t\}; \mathbf{k})$. For $c := \|f - g\|$, we have the inequalities

$$g - 2c \leq f - c \leq g \quad \text{and} \quad f - 2c \leq g - c \leq f.$$

The inclusions of sublevel sets

$$\{f < t\} \subset \{g < t + c\} \subset \{f < t + 2c\}$$

yield the desired interleaving relations. In other words, $d_{\text{int}}(V(f), V(g)) \leq \|f - g\|$.

The direct computation of d_{int} is almost impossible, since we need to deal with an $\mathbb{R}$-family of morphisms. Remarkably, this interleaving relation (as continuous data) can be translated into discrete data. Then the computation can be carried out in a more efficient way. The crux of this translation lies in the following definition.

Definition 2.15 Two barcodes $\mathcal{B}$ and C are δ-matched for a given $\delta > 0$ if, after erasing *some* intervals of length $< 2\delta$ in $\mathcal{B}$ and C, the rest can be matched in a one-to-one manner

$$(a, b] \in \mathcal{B} \quad \longleftrightarrow \quad (c, d] \in C$$

so that $|a - c| < \delta$ and $|b - d| < \delta$. Moreover, the bottleneck distance d_{bottle} is defined by

$$d_{\text{bottle}}(\mathcal{B}, C) = \inf\{\delta > 0 \mid \mathcal{B} \text{ and } C \text{ are } \delta\text{-matched}\}.$$

Here is another important theorem in the theory of persistence **k**-modules.

Theorem 2.7 (Isometry Theorem) *For any two persistence* **k**-*modules* V, W, $d_{\text{int}}(V, W) = d_{\text{bottle}}(\mathcal{B}(V), \mathcal{B}(W))$.

The proof of this theorem is difficult. The standard references for the proof are [8, 11] and [9] where an interpolation method is used, [3] where an elementary proof is provided, and [6] where a categorical perspective is investigated.

Remark 2.9 Note that for any $\phi \in \text{Diff}(M)$, we have $V(f) \simeq V(\phi^* f)$. Then by Example 2.14 above,

$$d_{\text{int}}(V(f), V(g)) \leq \inf_{\phi \in \text{Diff}(M)} \|f - \phi^* g\|. \tag{2.17}$$

A direct application of (2.17) is to answer the following question: roughly represented by Figure 2.1[7], how well can we approximate a C^0-function by a Morse function with exactly two critical points?

Exercise 2.12 In Figure 2.1, check that the answer to the question raised in Remark 2.9 is $\frac{a_3 - a_2}{2}$. (Hint: Use barcodes.)

2.8 Definition of the Singular Support

For the sake of brevity, denote $\mathcal{D}(\text{Sh}(\mathbf{k}_X))$ the derived category of sheaves of **k**-modules over the space X, by $\mathcal{D}(\mathbf{k}_X)$. The singular support of a sheaf $\mathcal{F} \in \mathcal{D}(\mathbf{k}_X)$ can be regarded as a geometrization of $\mathcal{F}$, which provides more information than just the support of $\mathcal{F}$. We first introduce the singular support of a sheaf $\mathcal{F} \in \text{Sh}(\mathbf{k}_X)$ in Definition 2.16, then upgrade it to $\mathcal{D}(\mathbf{k}_X)$ in Definition 2.17 (see Section 5.1 in [32]).

---

[7]This picture is borrowed from Section 4.2 in [43].

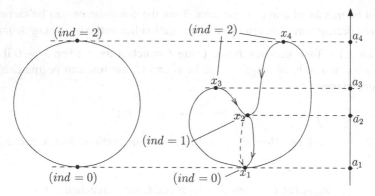

Fig. 2.1 C^0-approximation by a Morse function

Definition 2.16 Let $\mathcal{F} \in \mathrm{Sh}(\mathbf{k}_X)$. We say $\mathcal{F}$ *propagates* at $(x, \xi) \in T^*X$, if for any $\phi : M \longrightarrow \mathbb{R}$ with $\phi(x) = 0$ and $d\phi(x) = \xi$, one has

$$\lim_{\substack{\longrightarrow \\ x \in U}} H^*(\{\phi < 0\} \cup U; \mathcal{F}) \longrightarrow H^*(\{\phi < 0\}; \mathcal{F})$$

is an isomorphism. Then

$$SS(\mathcal{F}) := \overline{\{(x, \xi) \in T^*X \mid \mathcal{F} \text{ does } not \text{ propagate at } (x, \xi)\}}.$$

Geometrically, $(x, \xi) \notin SS(\mathcal{F})$ means that locally moving along the direction ξ at point $x \in X$ does not change the sheaf cohomology (so any section can be extended locally). There are some easy facts following directly from Definition 2.16. For instance, $SS(\mathcal{F})$ is always closed and conical (meaning $\mathbb{R}_+$-invariant in T^*X, where $\mathbb{R}_+$ acts on the co-vector components of elements in T^*X); $SS(\mathcal{F}) \cap 0_M = \mathrm{supp}(\mathcal{F})$. Also from this definition, one can check that

Example 2.15 $SS(\mathbf{k}_X) = 0_X$.

Example 2.16 (1) If D is a closed domain, then $SS(\mathbf{k}_D) = \nu^*_-(\partial D) \cup 0_D$. (2) If U is an open domain, then $SS(\mathbf{k}_U) = \nu^*_+(\partial \bar{U}) \cup 0_{\bar{U}}$. Here $\nu^*_-(\partial D)$ denotes the negative conormal bundle of ∂D - pointing inside; $\nu^*_+(\partial \bar{U})$ denotes the positive conormal bundle over $\partial \bar{U}$ - pointing outside. See Figure 2.2.

The following example, especially the details of its computation, is very useful.

Example 2.17 For $\mathbf{k}_{[a,b)} \in \mathrm{Sh}(\mathbf{k}_\mathbb{R})$,

$$SS(\mathbf{k}_{[a,b)}) = 0_{[a,b]} \cup (\{a\} \times \mathbb{R}_{\geq 0}) \cup (\{b\} \times \mathbb{R}_{\geq 0}).$$

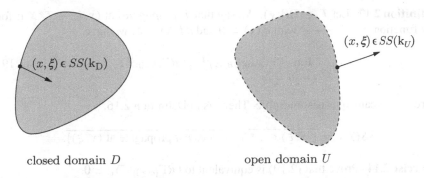

$(x, \xi) \in SS(\mathbf{k}_D)$

$(x, \xi) \in SS(\mathbf{k}_U)$

closed domain D open domain U

Fig. 2.2 SS of the constant sheaf over closed/open domain

In fact, at $x = b$, we can take $\phi(x) = x - b$ (so $\phi(b) = 0$ and $d\phi(b) = 1$). On the one hand,

$$H^*(\{\phi < 0\}; \mathbf{k}_{[a,b)}) = H^*((-\infty, b); \mathbf{k}_{[a,b)}) = \mathbf{k},$$

while on the other

$$\varinjlim_{x \in U} H^*(\{\phi < 0\} \cup U; \mathcal{F}) = \varinjlim_{\epsilon \longrightarrow 0} H^*((-\infty, b + \epsilon), \mathbf{k}_{[a,b)}) = 0.$$

In other words, at the point $(b, 1) \in T^*\mathbb{R}$ sheaf cohomology changes. Thus $\mathbf{k}_{[a,b)}$ does *not* propagate at $(b, 1)$ (and this holds for all (b, ξ) with $\xi > 0$). Similar argument works for $x = a$. For $x \in (a, b)$, it is trivial to see that sheaf cohomology will not change for any co-vector $\xi > 0$, so we are left with the case $(x, 0)$ where the co-vector is 0 for $x \in (a, b)$. Then taking $\phi(x) \equiv 0$ one has

$$H^*(\{\phi < 0\}; \mathbf{k}_{[a,b)}) = H^*(\emptyset; \mathbf{k}_{[a,b)}) = 0,$$

but

$$\varinjlim_{x \in U} H^*(\{\phi < 0\} \cup U; \mathcal{F}) = \varinjlim_{\epsilon \longrightarrow 0} H^*((x - \epsilon, x + \epsilon), \mathbf{k}_{[a,b)}) = \mathbf{k} \; (= \text{stalk}).$$

Thus $[a, b) \times \{0\}$ is also included in the singular support.

Exercise 2.13 Compute $SS(\mathbf{k}_{(a,b]})$, $SS(\mathbf{k}_{(a,b)})$, and $SS(\mathbf{k}_{[a,b]})$.

Now, let us update Definition 2.16 to be defined for $\mathcal{F} \in \mathcal{D}(\mathbf{k}_X)$. Recall that if $U \subset X$ is an open subset and $\Gamma_U : \mathrm{Sh}(\mathbf{k}_X) \longrightarrow \mathrm{Mod}_\mathbf{k}$ (the category of $\mathbf{k}$-modules) is the left exact functor defined by $\Gamma_U(\mathcal{F}) = \mathcal{F}(U)$. This is a (local) generalization of (1) in Example 2.1. Its derived functor is denoted by $R\Gamma_U$, and the i-th derived functor is

$$R^i \Gamma_U(\mathcal{F}) = h^i(R\Gamma_U \mathcal{F}) = H^i(U; \mathcal{F}). \qquad (2.18)$$

Definition 2.17 Let $\mathcal{F} \in \mathcal{D}(\mathbf{k}_X)$. We say that $\mathcal{F}$ *propagates* at $(x, \xi) \in T^*X$ if for any function $\phi : X \longrightarrow \mathbb{R}$ with $\phi(x) = 0$ and $d\phi(x) = \xi$, we have

$$\varinjlim_{\epsilon \longrightarrow 0} R\Gamma_{\{\phi<0\}\cup B_\epsilon(x)}\mathcal{F} \simeq R\Gamma_{\{\phi<0\}}\mathcal{F}. \tag{2.19}$$

Here "$\simeq$" means quasi-isomorphic. Then, as in Definition 2.16,

$$SS(\mathcal{F}) := \overline{\{(x, \xi) \in T^*X \mid \mathcal{F} \text{ does } not \text{ propagate at } (x, \xi)\}}.$$

Exercise 2.14 Prove that (2.19) is equivalent to $(R\Gamma_{\{\phi\geq0\}}\mathcal{F})_x \simeq 0$.

Proposition 2.5 *Given $\mathcal{F}_1, \mathcal{F}_2, \mathcal{F}_3 \in \mathcal{D}(\mathbf{k}_X)$ and a distinguished triangle $\mathcal{F}_1 \longrightarrow \mathcal{F}_2 \longrightarrow \mathcal{F}_3 \xrightarrow{+1}$, one has for $\{i, j, k\} = \{1, 2, 3\}$,*

$$SS(\mathcal{F}_k) \subset SS(\mathcal{F}_i) \cup SS(\mathcal{F}_j) \quad and \quad SS(\mathcal{F}_i)\triangle SS(\mathcal{F}_j) \subset SS(\mathcal{F}_k), \tag{2.20}$$

where $\triangle$ denotes the symmetric difference.

Proof Use the Five Lemma.   □

Example 2.18 If $a < b < c$, we have a distinguished triangle $\mathbf{k}_{[a,b)} \longrightarrow \mathbf{k}_{[a,c)} \longrightarrow \mathbf{k}_{[b,c)} \xrightarrow{+1}$. Then Proposition 2.5 says that

$$SS(\mathbf{k}_{[a,c)}) \subset SS(\mathbf{k}_{[a,b)}) \cup SS(\mathbf{k}_{[b,c)}).$$

Indeed, based on Example 2.17, we can see this directly from Figure 2.3. Moreover, this example also shows that in general the inclusions in (2.20) are strict.

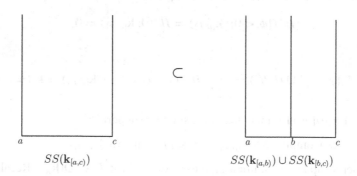

$$SS(\mathbf{k}_{[a,c)}) \qquad\qquad\qquad SS(\mathbf{k}_{[a,b)}) \cup SS(\mathbf{k}_{[b,c)})$$

Fig. 2.3 A triangle inequality for SS

2.9   Properties of the Singular Support

The singular support enjoys many interesting properties. The standard reference for all these properties is [32]. Here, we list (without proofs) some of these properties in three families. One concerns geometric properties (see Theorem 2.8), another concerns functorial properties (see Proposition 2.6, 2.7, 2.8), and one concerns the interesting feature that restrictions on the singular support can impose a strong constraint on the behavior of sheaves (see Lemma 2.3).

2.9.1   Geometric and Functorial Properties

Theorem 2.8 (Theorem 6.5.4 in [32]) *For any* $\mathcal{F} \in \mathcal{D}(\mathbf{k}_X)$, $SS(\mathcal{F})$ *is coisotropic.*

Remark 2.10 (1) Since $SS(\mathcal{F})$ can be highly degenerate, the rigorous concept of coisotropic is given by Definition 6.5.1 in [32] and is called *involutive*. For the smooth part, this coincides with the standard notion of coisotropic in the symplectic manifold $(T^*M, \omega_{\mathrm{can}})$. (2) In this book, most of the singular supports we will encounter are actually singular Lagrangian submanifolds (because most of the sheaves we will work with are constructible, i.e., there exists a stratification such that the restriction to each stratum is locally constant).

Various functors, say Rf_*, $Rf_!$, f^{-1}, $\otimes$, $R\mathrm{Hom}$, induce functorial properties on singular supports. Here we list the resulting consequences in coordinates.

Proposition 2.6 (Pushforward) *Suppose that* $f : X \longrightarrow Y$ *is proper on the support of* $\mathcal{F} \in \mathcal{D}(\mathbf{k}_X)$. *Then*

$$SS(Rf_*\mathcal{F}) \subset \big\{(y, \eta) \in T^*Y \mid \exists (x, \xi) \in SS(\mathcal{F}) \text{ s.t. } f(x) = y \text{ and } f^*(\eta) = \xi\big\}.$$

The same inclusion holds also for $Rf_!$.

Example 2.19 Suppose that $X = S^1$ and $Y = \{\mathrm{pt}\}$. Take $\mathcal{F}$ to be a locally constant but non-constant sheaf on S^1. Then $Rf_*(\mathcal{F}) = 0$ (**Exercise**) and so $SS(Rf_*(\mathcal{F})) = \emptyset$. This shows that the inclusion relation in Proposition 2.6 is in general strict. By Proposition 5.4.4. in [32], this inclusion is an equality if the smooth map $f : X \longrightarrow Y$ is a closed embedding.

Example 2.20 Let $X = M \times \mathbb{R}$, $Y = \mathbb{R}$. Let $f = \pi$ be the projection onto the $\mathbb{R}$-component. Then it is easy to check that $f^*(\eta) = (0, \eta)$. Therefore, for any $\mathcal{F} \in \mathcal{D}(\mathbf{k}_X)$,

$$SS(Rf_*\mathcal{F}) \subset \{(r, \eta) \in T^*\mathbb{R} \mid \exists (x, 0, r, \eta) \in SS(\mathcal{F})\}.$$

Let $f : X \longrightarrow Y$ be a morphism and let Λ be an $\mathbb{R}_+$-conic subset of T^*Y. We say f is *non-characteristic* (see Definition 5.4.12 in [32]) for Λ if for any point

$(y, \eta) \in \Lambda$ such that $y = f(x)$ for some $x \in X$, if $\eta(df_x(v)) = 0$ for all $v \in T_x X$, then $\eta = 0$. In particular, if f is a submersion, then f is non-characteristic for any $\mathbb{R}_+$-conic subset $\Lambda \subset T^* Y$.

Proposition 2.7 (Pullback) *Suppose that* $f : X \longrightarrow Y$ *is non-characteristic for* $SS(\mathcal{F})$ *for* $\mathcal{F} \in \mathcal{D}(\mathbf{k}_Y)$. *Then*

$$SS(f^{-1}\mathcal{F}) = \{(x, \xi) \in T^* X \mid \exists (y, \eta) \in SS(\mathcal{F}) \text{ s.t. } f(x) = y \text{ and } f^*(\eta) = \xi\}.$$

Example 2.21 Let $X = \mathbb{R} \times \mathbb{R}, Y = \mathbb{R}$ and let f be the summation map, i.e., $f(t_1, t_2) = t_1 + t_2$. Then it is easy to check that $f^*(\eta) = (\eta, \eta)$. Therefore, for any $\mathcal{F} \in \mathcal{D}(\mathbf{k}_Y)$,

$$SS(f^{-1}\mathcal{F}) = \{(t_1, \xi, t_2, \xi) \mid \exists (t_1 + t_2, \xi) \in SS(\mathcal{F})\}.$$

Exercise 2.15 Let $X = \mathbb{R}, Y = \mathbb{R} \times \mathbb{R}$ and let f be the diagonal embedding, i.e., $f(t) = (t, t)$. Write down the formula for $SS(f^{-1}\mathcal{F})$ for any $\mathcal{F} \in \mathcal{D}(\mathbf{k}_Y)$.

Proposition 2.8 (Tensor product and Hom) *For any* $\mathcal{F}, \mathcal{G} \in \mathcal{D}(\mathbf{k}_X)$,

(1) *if* $SS(\mathcal{F}) \cap SS(\mathcal{G})^a \subset 0_X$, *then*

$$SS(\mathcal{F} \otimes \mathcal{G}) \subset \{(x, \xi_1 + \xi_2) \in T^* X \mid (x, \xi_1) \in SS(\mathcal{F}) \text{ and } (x, \xi_2) \in SS(\mathcal{G})\};$$

(2) *if* $SS(\mathcal{F}) \cap SS(\mathcal{G}) \subset 0_X$, *then*

$$SS(R\mathcal{H}om(\mathcal{F}, \mathcal{G})) \subset \{(x, -\xi_1 + \xi_2) \in T^* X \mid (x, \xi_1) \in SS(\mathcal{F}) \text{ and } (x, \xi_2) \in SS(\mathcal{G})\},$$

where in (1), *"a" means taking the negative of the co-vector part.*

Corollary 2.1 (External tensor product) *Let* $\mathcal{F} \in \mathcal{D}(\mathbf{k}_X)$ *and* $\mathcal{G} \in \mathcal{D}(\mathbf{k}_Y)$.

$$SS(\mathcal{F} \boxtimes \mathcal{G}) \subset \{(x, \xi, y, \eta) \in T^*(X \times Y) \mid (x, \xi) \in T^* X, \ (y, \eta) \in T^* Y\}.$$

Proof By definition, $\mathcal{F} \boxtimes \mathcal{G} = (p_1^{-1}\mathcal{F}) \otimes (p_2^{-1}\mathcal{G})$, where p_1 and p_2 are the projections from $X \times Y$ onto X and onto Y, respectively. Then, by Proposition 2.7,

$$SS(p_1^{-1}\mathcal{F}) = \{(x, \xi, y, 0) \in T^*(X \times Y) \mid (x, \xi) \in SS(\mathcal{F})\},$$

$$SS(p_2^{-1}\mathcal{G}) = \{(x, 0, y, \eta) \in T^*(X \times Y) \mid (y, \eta) \in SS(\mathcal{G})\}.$$

Note that $SS(p_1^{-1}\mathcal{F}) \cap SS(p_2^{-1}\mathcal{G})^a \subset 0_{X \times Y}$, therefore by (1) in Proposition 2.8,

$$SS((p_1^{-1}\mathcal{F}) \otimes (p_2^{-1}\mathcal{G})) \subset \{(x, \xi, y, \eta) \in T^*(X \times Y) \mid (x, \xi) \in T^* X, \ (y, \eta) \in T^* Y\}.$$

This completes the proof. $\qquad\qquad\qquad\qquad\qquad\qquad\qquad\qquad\qquad\qquad\qquad\qquad$ $\square$

2.9.2 Microlocal Morse Lemma

Here is a useful (but non-trivial) observation.

Lemma 2.3 *If $\mathcal{F}$ is a complex of sheaves on X and $SS(\mathcal{F}) \subset 0_X$, then $\mathcal{F}$ is quasi-isomorphic to a locally constant sheaf.*

To rigorously prove this lemma, we need the important *microlocal Morse lemma* (see Corollary 5.4.19 in [32]). A low-tech formulation goes as follows.

Theorem 2.9 *Let $\mathcal{F} \in \mathcal{D}(\mathbf{k}_{\mathbb{R}})$ with compact support. If $SS(\mathcal{F})|_{[a,b]}$ has only non-positive co-vectors, then*

$$H^*((-\infty, a); \mathcal{F}) \simeq H^*((-\infty, b); \mathcal{F}).$$

In other words, sections of cohomology can propagate from a to b.

Example 2.22 Here is an example illustrating Theorem 2.9 for degree $* = 0$. Checking higher degrees can be complicated in general. For $\mathcal{F} = \mathbf{k}_{(0,1]}$, an elementary computation similar to Example 2.17 shows that $SS(\mathcal{F})$ satisfies the assumption in Theorem 2.9. On the other hand, by the definition of $\mathbf{k}_{(0,1]}$,

$$H^0((-\infty, a); \mathcal{F}) = H^0((-\infty, b); \mathcal{F}) = H^0((-\infty, c); \mathcal{F}) = 0$$

for any $a < 0 < b < 1 < c$.

Theorem 2.9 implies a geometric formulation of microlocal Morse lemma.

Theorem 2.10 (Geometric version of microlocal Morse lemma) *Let $\mathcal{F} \in \mathcal{D}(\mathbf{k}_X)$ with compact support. Let $f : X \longrightarrow \mathbb{R}$ be a differentiable function, and suppose that $H^*(X; \mathcal{F}) \neq 0$. Then $SS(\mathcal{F}) \cap \mathrm{graph}(df) \neq \emptyset$.*

Example 2.23 Let $\mathcal{F} = \mathbf{k}_X$, the constant sheaf on X. Since $SS(\mathcal{F}) = 0_X$, Theorem 2.10 reduces to the classical Morse lemma, i.e., non-vanishing of (Morse) cohomology detects critical points.

Proof (of Theorem 2.10 assuming Theorem 2.9) We will prove the counter-positive, i.e., if $SS(\mathcal{F}) \cap \mathrm{graph}(df) = \emptyset$, then $H^*(X; \mathcal{F}) = 0$. First of all, we know that $H^*(f^{-1}(-\infty, a); \mathcal{F}) \simeq H^*((-\infty, a); Rf_*\mathcal{F})$ for any $a \in \mathbb{R}$. By the pushforward formula for SS (see Proposition 2.6),

$$SS(Rf_*\mathcal{F}) \subset \left\{ (y, \eta) \in T^*\mathbb{R} \, \middle| \, \begin{matrix} \exists x \in X \text{ s.t. } f(x) = y \\ (x, f^*\eta) \in SS(\mathcal{F}) \end{matrix} \right\} (:= \Lambda_f(SS(\mathcal{F}))).$$

$$\tag{2.21}$$

In particular, for $f : X \longrightarrow \mathbb{R}$, $f^*\eta$ can be expressed as $\eta \cdot df(x)$ where η is identified with a number (possibly 0). Our assumption implies that, for any $\eta > 0$, $(y, \eta) \notin \Lambda_f(SS(\mathcal{F}))$ for all $y \in [a, b]$. Then, by (2.21),

$$SS(Rf_*\mathcal{F}) \cap T^*\mathbb{R}|_{[a,b]} \subset \Lambda_f(SS(\mathcal{F})) \cap T^*\mathbb{R}|_{[a,b]}$$

$$\subset \{(y, \eta) \in T^*\mathbb{R}|_{[a,b]} \mid \eta \leq 0\}.$$

Now Theorem 2.9 implies that

$$H^*(f^{-1}(-\infty, a); \mathcal{F}) \simeq H^*(f^{-1}(-\infty, b); \mathcal{F}).$$

Finally, thanks to the assumption that supp$(\mathcal{F})$ is compact, we can take $a << 0$ and $b >> 0$ such that $H^*(f^{-1}(-\infty, a); \mathcal{F}) = 0$ and $H^*(f^{-1}(-\infty, b); \mathcal{F}) = H^*(X; \mathcal{F})$, and then we get the desired conclusion.   □

To end this section, let us give the proof of Lemma 2.3.

Proof For each $x \in X$, consider the differentiable function on X defined by $f(y) = d_\rho(y, x)^2$ under some fixed metric ρ on X. Since the only critical point of f is x itself, by Theorem 2.10, one knows that, for some $\epsilon > 0$, $H^*(B(x, \epsilon); \mathcal{F}) \simeq H^*(B(x, \epsilon'); \mathcal{F})$ for any $0 < \epsilon' \leq \epsilon$. In other words, $R\Gamma(B(x, \epsilon); \mathcal{F}) \simeq R\Gamma(\mathcal{F})_x$, that is, $R\Gamma(\mathcal{F})$ or, equivalently, the cohomology sheaves $H^*(\mathcal{F})$ for all degree $* \in \mathbb{Z}$ are locally constant.   □

Remark 2.11 The proof of Theorem 2.9 is essentially subtler than its statement may suggest (the deep reason being that the projective/inverse limit is not exact, so it does not necessarily commute with taking cohomology). More explicitly, we need to compare $H^*((-\infty, b); \mathcal{F})$ and $\varprojlim_{a<b} H^*((-\infty, a); \mathcal{F})$. The complete proof of Theorem 2.9 is based on a sophisticated argument called "Mittag-Leffler inductive procedure" (see Section 1.12 in [32]).

Last but not least, under a transversality assumption, we have a finer statement of microlocal Morse lemma, called *microlocal Morse inequality*. Recall the classical version of Morse inequality. For a Morse function $f : X \longrightarrow \mathbb{R}$, denote $c_j(f, X) = $ #{critical points of index j} and $b_j(X) = \dim_\mathbf{k} H^j(X; \mathbf{k})$. Then for any $l \in \mathbb{N} \cup \{0\}$,

$$\sum_{j=0}^{l}(-1)^{l+j}b_j(X) \leq \sum_{j=0}^{l}(-1)^{l+j}c_j(f, X).$$

In particular, for any $j \in \mathbb{N} \cup \{0\}$, $b_j(\mathcal{F}) \leq c_j(f, X)$ (this is usually called the *weak* Morse inequality). Here are two more concepts.

- For a given sheaf (of $\mathbf{k}$-module) $\mathcal{F}$ on a manifold X, define "the sheaf version of Betti number" as follows: for each $j \in \mathbb{N} \cup \{0\}$, $b_j(\mathcal{F}) := \dim_\mathbf{k} H^j(X; \mathcal{F})$.

- For $(x, p) \in SS(\mathcal{F})$ (with the test function ϕ specified in concrete situations), denote $V_x(\phi) := \{(R^*\Gamma_{\{\phi \geq 0\}}(\mathcal{F}))_x\}_{* \in \mathbb{Z}}$[8] and for $j \in \mathbb{N} \cup \{0\}$, define $b_j(V_x(\phi)) = \dim_{\mathbf{k}}(R^j\Gamma_{\{\phi \geq 0\}}(\mathcal{F}))_x$.

Now we can state microlocal Morse inequality.

Theorem 2.11 (Proposition 5.4.20 **in** [32]) *Let $\mathcal{F}$ be a sheaf with compact support on a manifold X such that for any $j \in \mathbb{N} \cup \{0\}$, $b_j(\mathcal{F}) < \infty$. Suppose $f : X \longrightarrow \mathbb{R}$ is a C^1-function such that*

$$\mathrm{graph}(df) \cap SS(\mathcal{F}) = \{(x_1, p_1), \ldots, (x_N, p_N)\},$$

and assume that each $V_{x_i}(\phi_i)$ is finitely indexed and finite-dimensional for each degree $j \in \mathbb{N} \cup \{0\}$, where $\phi_i(x) = f(x) - f(x_i)$. Then for any $l \in \mathbb{N} \cup \{0\}$,

$$\sum_{j=0}^{l}(-1)^{l+j}b_j(\mathcal{F}) \leq \sum_{i=1}^{N}\sum_{j=0}^{l}(-1)^{l+j}b_j(V_{x_i}(\phi_i)). \tag{2.22}$$

In particular, for any $j \in \mathbb{N} \cup \{0\}$, $b_j(\mathcal{F}) \leq \sum_{i=1}^{N}b_j(V_{x_i}(\phi_i))$.

Example 2.24 Let $\mathcal{F} = \mathbf{k}_X$. Then $b_j(\mathcal{F}) = b_j(X)$ (the classical Betti number). We can check (**Exercise**) that

$$b_j(V_{x_i}(\phi_i)) = \begin{cases} i, & \text{if } j = \text{Morse index of } x_i, \\ 0, & \text{otherwise.} \end{cases}$$

In other words, this recovers the classical Morse inequality. For a specific case take $X = \mathbb{R}^2, \mathcal{F} = \mathbf{k}_X$ and $f(x, y) = x^2 - y^2$. One can compute that $b_1(V_{(0,0)}(f)) = 1$, and $b_i(V_{(0,0)}(f)) = 0$ for degrees 0 and 2.

---

[8]This is a collection of graded vector spaces. By the definition of $SS(\mathcal{F})$, if $(x, p) \in SS(\mathcal{F})$, the collection $V_x(\phi)$ is non-zero for certain degrees.

Chapter 3
Tamarkin Category Theory

Abstract In this chapter, we define and then study in detail Tamarkin categories. A Tamarkin category is defined as a categorical orthogonal complement, and the elements in a Tamarkin category can be completely characterized by a sheaf operator - sheaf convolution. A restrictive version of Tamarkin category can capture the geometry of subsets of the base manifold. There is a section specially devoted to illustrating this when the subsets are Lagrangian submanifolds. A fundamental result on this restrictive Tamarkin category is the Separation Theorem, which can be viewed as a microlocal generalization of the straightforward observation that the set of morphisms between two sheaves with disjoint supports is trivial. The proof of the Separation Theorem is provided once the adjoint sheaf is introduced, and this theorem will be important in the discussion of displacement energy in later chapters. Finally, inspired by the theory of persistence **k**-modules, an interleaving distance in a Tamarkin category is defined, and a purely algebraic criterion to check interleaving relations is provided. This criterion is handy in the sense that it boils down to computations of singular supports.

3.1 Categorical Orthogonal Complement

In this section, we introduce the orthogonality in categories, which plays a crucial role in the construction of Tamarkin categories. Recall that in linear algebra, due to the naturally defined inner product, we can define the orthogonal complement of a subspace in a fixed vector space. In the case of a category, with the help of $\mathrm{Hom}(\cdot, \cdot)$, we can define a certain orthogonal complement of a subcategory in a fixed category.

Definition 3.1 Let C be a category and C' be a (full) subcategory of C. Define the *left orthogonal complement* of C' in C by

$$(C')^{\perp, l} := \{x \in C \mid \mathrm{Hom}_C(x, y) = 0 \text{ for any } y \in C'\},$$

and the right orthogonal complement of C' in C by

$$(C')^{\perp, r} := \{x \in C \mid \mathrm{Hom}_C(y, x) = 0 \text{ for any } y \in C'\}.$$

© Springer Nature Switzerland AG 2020

J. Zhang, *Quantitative Tamarkin Theory*, CRM Short Courses,
https://doi.org/10.1007/978-3-030-37888-2_3

Remark 3.1 Note that both $(C')^{\perp,l}$ and $(C')^{\perp,r}$ are subcategories of C.

Example 3.1 Let $\mathcal{A}$ be the category of finitely generated abelian groups, and let the subcategory $\mathcal{A}'$ consist of the finitely generated abelian torsion groups. Then

$$(\mathcal{A}')^{\perp,l} = \{A \in \mathcal{A} \mid \text{Hom}(A, B) = 0 \text{ for any } B \in \mathcal{A}'\} = \{0\}.$$

In fact, there always exist non-trivial morphisms from any finitely generated group to some torsion group. However, the only morphism from a torsion group to any free group is just zero. Therefore, one gets

$$(\mathcal{A}')^{\perp,r} = \{A \in \mathcal{A} \mid \text{Hom}(B, A) = 0 \text{ for any } B \in \mathcal{A}'\} = \{A \in \mathcal{A} \mid A \text{ is free}\}.$$

Note that this example also shows that the left orthogonal complement and the right orthogonal complement are in general not the same.

Example 3.2 Let $\mathcal{P}$ be the category of persistence **k**-modules and let $\mathcal{P}'$ be the subcategory consisting of the "torsion" persistence **k**-modules, i.e., those for which the corresponding barcodes only have bars of finite length. Then, similarly to Example 3.1,

$$(\mathcal{P}')^{\perp,l} = \{V \in \mathcal{P} \mid \text{Hom}(V, W) = 0 \text{ for any } W \in \mathcal{P}'\} = \{0\}.$$

For instance, there exists non-trivial morphisms from $\mathbb{I}_{(0,\infty)}$ to $\mathbb{I}_{(1,0]}$. However, any morphism from $\mathbb{I}_{(a,b]}$ with $b < \infty$ to $\mathbb{I}_{(c,\infty)}$ is zero. Therefore, we get

$$(\mathcal{P}')^{\perp,r} = \{V \in \mathcal{P} \mid \text{Hom}(W, V) = 0 \text{ for any } W \in \mathcal{P}'\}$$
$$= \{V \in \mathcal{P} \mid \mathcal{B}(V) \text{ consists of only infinite length bars}\}.$$

Remark 3.2 (Exercise) If $\mathcal{T}'$ is a triangulated subcategory of a triangulated category $\mathcal{T}$, then its left/right orthogonal complement is also a triangulated subcategory. Due to this exercise, it seems plausible that both $\mathcal{A}$ and $\mathcal{P}$ *cannot* be viewed as triangulated categories. For instance, $\mathbb{Z} \xrightarrow{\times 2} \mathbb{Z}$ cannot be completed as a distinguished triangle inside $(\mathcal{A}')^{\perp,r}$. On the other hand, since $\mathcal{P}$ has the special property that it has homological dimension ≤ 1, there is a chance to upgrade $\mathcal{P}$ to a triangulated category once we add the "flavor" of derived category (see Sect. 3.10).

Orthogonal complements are closely related to adjoint functors. Let us explain this.

Proposition 3.1 *Let C be a derived category and C' be a subcategory of C. The inclusion $i : C' \to C$ has a left adjoint $p : C \to C'$ if and only if for any $x \in C$, there exists a distinguished triangle $z \to x \to y \xrightarrow{+1}$ such that $z \in (C')^{\perp,l}$ and $y \in C'$.*

Proof Suppose that i has a left adjoint p. For any $x \in C$, define $y := p(x) \in C'$. Then by item (0) of Proposition 2.2, there exists a distinguished triangle

$$x \longrightarrow y \longrightarrow z[1] \xrightarrow{+1}, \quad \text{for some } z[1] \in C.$$

This is the same as $z \to x \to y \xrightarrow{+1}$. We just need to check that $z \in (C')^{\perp,\mathrm{l}}$. In fact, for any $w \in C'$, applying $\mathrm{Hom}(\cdot, w)$ to the distinguished triangle above one gets by Lemma 3.6, a long exact sequence

$$\mathrm{Hom}(y, w)(= \mathrm{Hom}(p(x), w)) \longrightarrow \mathrm{Hom}(x, w) \longrightarrow \mathrm{Hom}(z, w) \xrightarrow{+1} .$$

By the adjoint relation, $\mathrm{Hom}(p(x), w) = \mathrm{Hom}(x, i(w)) = \mathrm{Hom}(x, w)$, which implies that $\mathrm{Hom}(z, w) = 0$. Conversely, assume the existence of the distinguished triangle in the statement. Define $p : C \to C'$ by $p(x) := y$. Then, for any $w \in C'$, applying $\mathrm{Hom}(\cdot, w)$, one gets the long exact sequence as above. Since $\mathrm{Hom}(z, w) = 0$, we get the adjoint relation $\mathrm{Hom}(p(x), w) = \mathrm{Hom}(x, w) = \mathrm{Hom}(x, i(w))$. $\square$

Remark 3.3 The same argument works for the right orthogonal complement, i.e., the existence of a right adjoint of inclusion $i : C' \to C$ is equivalent to the existence of a distinguished triangle $y \to x \to z \xrightarrow{+1}$ such $y \in C'$ and $z \in (C')^{\perp,\mathrm{r}}$.

Remark 3.4 To justify that the map $p : C \to C'$ defined by $p(x) := y$ in the proof of Proposition 3.1 is indeed a functor, we also need to specify how p acts on morphisms. This is slightly complicated, so we address it in this remark.

Let us show the following property first: given a distinguished triangle $z \xrightarrow{g} x \xrightarrow{h} y \xrightarrow{+1}$ (so $h \circ g = 0$) in C and a morphism $f : x \to y'$ for some $y' \in C$, there exists a morphism $s : y \to y'$ such that $f = s \circ h$ if and only if $f \circ g = 0$. One direction is trivial. If there exists such a morphism $s : y \to y'$ with the desired property, then $f \circ g = (s \circ h) \circ g = s \circ (h \circ g) = 0$. For the other direction, if $f \circ g = 0$, then we have the commutative diagram

$$
\begin{array}{ccccccc}
z & \xrightarrow{g} & x & \xrightarrow{h} & y & \xrightarrow{+1} & \\
\downarrow{\scriptstyle 0} & & \downarrow{\scriptstyle f} & & \downarrow{\scriptstyle s} & & \\
0 & \xrightarrow{} & y' & \xrightarrow{1} & y' & \xrightarrow{+1} &
\end{array}
$$

The existence of the morphism $s : y \to y'$ is guaranteed by one of the axioms of a triangulated category. Moreover, by the commutativity of this diagram, $f = s \circ h$.

Now, recall our hypothesis that for any $x \in C$, there exists a distinguished triangle $z \xrightarrow{g} x \xrightarrow{h} y \xrightarrow{+1}$ such that $z \in (C')^{\perp,\mathrm{l}}$ and $y \in C'$. For $x \in C$ and $f' \in \mathrm{Hom}(x, x')$, consider the diagram

$$z \xrightarrow{\;g\;} x \xrightarrow{\;h\;} y \xrightarrow{\;+1\;}$$

$$\Big\downarrow f' \qquad \Big| s$$

$$z' \xrightarrow{\;g'\;} x' \xrightarrow{\;h'\;} y' \xrightarrow{\;+1\;}$$

Let $f := h' \circ f'$. Then $f \circ g = 0$ since $z \in\in (C')^{\perp,1}$ and $y' \in C'$. Hence, by the property above, we know that there exists a morphism $s \in \mathrm{Hom}(y, y')$ such that the diagram commutes. One can even show that s is the unique morphism to make this diagram commute. Then our desired action $p : \mathrm{Hom}(x, x') \to \mathrm{Hom}(y, y')$ is defined by $p(f') = s$.

Corollary 3.1 *Let C be a derived category and let C' be a subcategory of C. If the inclusion $i : C' \to C$ admits a left adjoint $p : C \to C'$, then $p(v) = 0$ for any $v \in (C')^{\perp,1}$.*

Proof Suppose not, that is, there exists some $v \in (C')^{\perp,1}$ such that $p(v) \neq 0$. By Proposition 3.1, there exists a distinguished triangle, $z \to v \to p(v) \xrightarrow{+1}$ such that $z \in (C')^{\perp,1}$ and $p(v) \in C'$. Applying $\mathrm{Hom}(\cdot, p(v))$ to this distinguished triangle, one gets the long exact sequence

$$\mathrm{Hom}(p(v), p(v)) \longrightarrow \mathrm{Hom}(v, p(v)) \longrightarrow \mathrm{Hom}(z, p(v)) \xrightarrow{+1} .$$

Note that $\mathrm{Hom}(z, p(v)) = 0$ implies $\mathrm{Hom}(v, p(v)) \simeq \mathrm{Hom}(p(v), p(v))$. Since $p(v) \neq 0$, the identity map $\mathbb{1}_{p(v)}$ corresponds to some *non-zero* map in $\mathrm{Hom}(v, p(v))$, which is a contradiction because $\mathrm{Hom}(v, p(v)) = 0$ (since $v \in (C')^{\perp,1}$). □

Remark 3.5 The proof of Corollary 3.1 given above was designed to demonstrate how to use Proposition 3.1. Certainly there exists a much shorter proof by using the adjoint relation. Explicitly, $\mathrm{Hom}(p(v), p(v)) = \mathrm{Hom}(v, ip(v)) = 0$, by the assumption that $v \in (C')^{\perp,1}$. Then we know that $p(v) = 0$ (otherwise $\mathbb{1}_{p(v)}$ will be a non-zero element in $\mathrm{Hom}(p(v), p(v))$).

3.2 Definitions of Tamarkin Categories

In this section, we give the definitions of Tamarkin categories. Recall that the category we are mainly working with is

$$\mathcal{D}(\mathbf{k}_X)(= \mathcal{D}(\mathrm{Sh}(\mathbf{k}_X))) := \text{the derived category of sheaves of } \mathbf{k}\text{-modules over } X,$$

whose objects are complexes of sheaves of $\mathbf{k}$-modules, denoted by $\mathcal{F}$. Also, recall that in a derived category quasi-morphisms are invertible. Moreover, $\mathcal{D}(\mathbf{k}_X)$ is a triangulated category. Let $X = M \times \mathbb{R}$, with the coordinate of $\mathbb{R}$ labelled t and the co-vector coordinate labelled τ. Consider the full subcategory of $\mathcal{D}(\mathbf{k}_X)$ defined as

$$\mathcal{D}_{\{\tau \leq 0\}}(\mathbf{k}_{M \times \mathbb{R}}) := \{\mathcal{F} \in \mathcal{D}(\mathbf{k}_{M \times \mathbb{R}}) \mid SS(\mathcal{F}) \subset \{\tau \leq 0\}\},$$

where $\{\tau \leq 0\}$ denotes the subset of $T^*(M \times \mathbb{R})$ where the τ-part is non-positive. Similarly, denote by $\{\tau > 0\}$ the subset where the τ-part is positive. Importantly, $\mathcal{D}_{\{\tau \leq 0\}}(\mathbf{k}_{M \times \mathbb{R}})$ is a triangulated subcategory. For instance, for $\mathcal{F} \to \mathcal{G}$, first complete it to be a distinguished triangle

$$\mathcal{F} \longrightarrow \mathcal{G} \longrightarrow \mathcal{H} \xrightarrow{+1} .$$

Then, by Proposition, 2.5, $SS(\mathcal{H}) \subset SS(\mathcal{F}) \cup SS(\mathcal{G}) \subset \{\tau \leq 0\}$. In other words, every distinguished triangle can be completed inside $\mathcal{D}_{\{\tau \leq 0\}}(\mathbf{k}_{M \times \mathbb{R}})$.

The role of the extra variable $\mathbb{R}$ may seem mysterious at the first sight, but in the following few sections we will see that $\mathbb{R}$ plays an highly important role in Tamarkin categories. There exists a well-defined reduction map $\rho : T^*_{\{\tau > 0\}}(M \times \mathbb{R}) \to T^*M$, given by

$$\rho(x, \xi, t, \tau) = (x, \xi/\tau) \tag{3.1}$$

Definition 3.2 (Definitions of Tamarkin categories) There are two versions of Tamarkin categories.

(1) (free version) Let

$$\mathcal{T}(M) := \mathcal{D}_{\{\tau \leq 0\}}(\mathbf{k}_{M \times \mathbb{R}})^{\perp, l}.$$

(2) (restricted version) For a given closed subset $A \subset T^*M$, let

$$\mathcal{T}_A(M) := \left\{ \mathcal{F} \in \mathcal{T}(M) \mid SS(\mathcal{F}) \subset \overline{\rho^{-1}(A)} \right\}$$

where the closure is taken in $T^*(M \times \mathbb{R})$.

Remark 3.6 Here we defined $\mathcal{T}(M)$ by using the *left* orthogonal complement. In fact, $\mathcal{T}(M)$ can also be defined as $\mathcal{D}_{\{\tau \leq 0\}}(\mathbf{k}_{M \times \mathbb{R}})^{\perp, r}$ via the *right* orthogonal complement. This is closely related to an operator called *adjoint sheaf*, defined in Sect. 3.7 and Corollary 3.3.

Example 3.3 The simplest example for $\mathcal{T}(M)$ is when $M = \{\text{pt}\}$, that is, the objects are complexes of sheaves over $\mathbb{R}$. For the sake of convenience, we will only consider constructible sheaves. For a decomposition theorem (where we use our hypothesis of constructibility) in [34], the typical element in $\mathcal{D}_{\{\tau \leq 0\}}(\mathbf{k}_{\mathbb{R}})$ is $\bigoplus \mathbf{k}_{(a,b]}[d]$, where d labels the degree shift of $\mathbf{k}_{(a,b]}$. Then a typical element in $\mathcal{T}(\text{pt})$ is $\bigoplus \mathbf{k}_{[c,d)}$, which can be identified with a persistence $\mathbf{k}$-module. See Appendix A.1 for a detailed explanation of this identification.

Example 3.4 The restricted version $\mathcal{T}_A(M)$ can be roughly divided into two cases. One is that A is a Lagrangian submanifold of T^*M, e.g., a Lagrangian submanifold which admits a generating function; the other is that A is a domain of T^*M, e.g., the complement of the standard open ball in $T^*\mathbb{R}^n (\simeq \mathbb{R}^{2n})$. The general philosophy is that when A is a Lagrangian submanifold, $\mathcal{T}_A(M)$ encodes the information of Lagrangian Floer homology; when A is a domain, $\mathcal{T}_A(M)$ encodes the information of symplectic homology.

Here, we give an easy example of the first case. Let $A = 0_M$, the zero-section of T^*M. Then,

$$\overline{\rho^{-1}(0_M)} = \{(m, 0, t, \tau) \mid m \in M, \ \tau \geq 0\}.$$

We claim that $\mathbf{k}_{M \times [0,\infty)} \in \mathcal{T}_{0_M}(M)$. First,

$$SS(\mathbf{k}_{M \times [0,\infty)}) = 0_M \times (\{(0, \tau) \mid \tau \geq 0\} \cup \{(t, 0) \mid t \geq 0\}) \subset \overline{\rho^{-1}(0_M)}.$$

The non-trivial part is to confirm that, for any $\mathcal{G} \in \mathcal{D}_{\{\tau \leq 0\}}(\mathbf{k}_{M \times \mathbb{R}})$, $\mathrm{Hom}(\mathbf{k}_{M \times [0,\infty)}, \mathcal{G}) = 0$. In fact, by the exact triangle $\mathbf{k}_{M \times (-\infty, 0)} \to \mathbf{k}_{M \times \mathbb{R}} \to \mathbf{k}_{M \times [0,\infty)} \xrightarrow{+1}$, we have the following computation in $\mathcal{T}(M)$:

$$\mathrm{RHom}(\mathbf{k}_{M \times [0,\infty)}, \mathcal{G}) = \mathrm{Cone}(\mathrm{RHom}(\mathbf{k}_{M \times \mathbb{R}}, \mathcal{G}) \to \mathrm{RHom}(\mathbf{k}_{M \times (-\infty, 0)}, \mathcal{G}))$$

$$= \mathrm{Cone}(R\Gamma(M \times \mathbb{R}, \mathcal{G}) \to R\Gamma(M \times (-\infty, 0), \mathcal{G})) = 0,$$

where the final step comes from the microlocal Morse lemma (see Theorem 2.9), due to the hypothesis on the singular support of $\mathcal{G}$.

Remark 3.7 Thanks to the functorial property of the singular support, Corollary 2.1, for any $\mathcal{F} \in \mathcal{T}(pt)$, $\mathbf{k}_M \boxtimes \mathcal{F} \in \mathcal{T}_{0_M}M$. In fact, locally in $\mathcal{T}_{0_M}M$, every element is of this form (see (ii) in Proposition 5.4.5 in [32]).

3.3 Sheaf Convolution and Composition

3.3.1 Definitions of Operators

For any $\mathcal{F}, \mathcal{G} \in \mathcal{D}(\mathbf{k}_{M \times \mathbb{R}})$, consider the diagram

$$
\begin{array}{ccc}
M \times M \times \mathbb{R} \times \mathbb{R} & \xrightarrow{\ \ s\ \ } M \times M \times \mathbb{R} \xrightarrow{\delta^{-1}} M \times \mathbb{R} \\
{}^{\pi_1}\swarrow \qquad \searrow^{\pi_2} & \\
M \times \mathbb{R} \qquad\qquad\qquad M \times \mathbb{R} &
\end{array}
$$

$$(3.2)$$

where

- $\pi_1(m_1, m_2, t_1, t_2) = (m_1, t_1)$;
- $\pi_2(m_1, m_2, t_1, t_2) = (m_2, t_2)$;
- $s(m_1, m_2, t_1, t_2) = (m_1, m_2, t_1 + t_2)$;
- $\delta(m, t) = (m, m, t)$.

Then one can define the following operator.

Definition 3.3 (Sheaf convolution)

$$\mathcal{F} * \mathcal{G} := \delta^{-1} Rs_!(\pi_1^{-1} \mathcal{F} \otimes \pi_2^{-1} \mathcal{G})(= \delta^{-1} Rs_!(\mathcal{F} \boxtimes \mathcal{G})).$$

Example 3.5 Let $\mathcal{F} \in \mathcal{D}(\mathbf{k}_{M \times \mathbb{R}})$. Then $\mathcal{F} * \mathbf{k}_{M \times \{0\}} = \mathcal{F}$.

Exercise 3.1 $\mathbf{k}_{(a,b)} * \mathbf{k}_{[0,\infty)} = \mathbf{k}_{[b,\infty)}[-1]$.

Example 3.6

$$\mathbf{k}_{[a,b)} * \mathbf{k}_{[c,d)} \simeq \begin{cases} \mathbf{k}_{[a+c,b+c)} \oplus \mathbf{k}_{[a+d,b+d)}[-1], & \text{if } b + c < a + d, \\ \mathbf{k}_{[a+c,a+d)} \oplus \mathbf{k}_{[b+c,b+d)}[-1], & \text{if } b + c \geq a + d. \end{cases}$$

Remark 3.8 (1) We also need another (more technical) *non-proper* convolution, defined by $\mathcal{F} *_{\mathrm{np}} \mathcal{G} = \delta^{-1} Rs_*(\pi_1^{-1} \mathcal{F} \otimes \pi_2^{-1} \mathcal{G})$. Sometimes this will change the results of computations. For instance, in Exercise 3.1, $\mathbf{k}_{(-\infty,b)} * \mathbf{k}_{[0,\infty)} = \mathbf{k}_{[b,\infty)}[-1]$. If we change to non-proper convolution,

$$\mathbf{k}_{(-\infty,b)} *_{\mathrm{np}} \mathbf{k}_{[0,\infty)} = \mathbf{k}_{(-\infty,b)}.$$

(2) Example 3.6 displays an interesting similarity to the formula for the tensor product of persistence $\mathbf{k}$-modules (cf. [45] and [62]).

Here is another useful sheaf operator.

Definition 3.4 (sheaf composition) For any $\mathcal{F} \in \mathcal{D}(\mathbf{k}_{X \times Y})$ and $\mathcal{G} \in \mathcal{D}(\mathbf{k}_{Y \times Z})$, define

$$\mathcal{F} \circ \mathcal{G} = R\pi_{3!}(\pi_1^{-1} \mathcal{F} \otimes \pi_2^{-1} \mathcal{G}),$$

where $\pi_1 : X \times Y \times Z \to X \times Y$, $\pi_2 : X \times Y \times Z \to Y \times Z$ and $\pi_3 : X \times Y \times Z \to X \times Z$ are projections.

Example 3.7 When $X = \{\mathrm{pt}\}$, any fixed $\mathcal{G} \in \mathcal{D}(\mathbf{k}_{Y \times Z})$ will serves as an operator (usually called kernel) $\circ Z : \mathcal{D}(\mathbf{k}_Y) \to \mathcal{D}(\mathbf{k}_Z)$.

Sometimes we can mix convolution and composition. The most general definition reads:

Definition 3.5 For any $\mathcal{F} \in \mathcal{D}(\mathbf{k}_{X \times Y \times \mathbb{R}})$ and $\mathcal{G} \in \mathcal{D}(\mathbf{k}_{Y \times Z \times \mathbb{R}})$, let

$$\mathcal{F} \bullet_Y \mathcal{G} = R p_{13!}(p_{12}^{-1}\mathcal{F} \otimes p_{23}^{-1}\mathcal{G}),$$

where $p_{13} : X \times Y \times Z \times \mathbb{R}^2 \to X \times Z \times \mathbb{R}$ by $(x, y, z, t_1, t_2) = (x, z, t_1 + t_2)$ and $p_{12} : X \times Y \times Z \times \mathbb{R}^2 \to X \times Y \times \mathbb{R}_1$ and $p_{23} : X \times Y \times Z \times \mathbb{R}^2 \to Y \times Z \times \mathbb{R}_2$ are projections. We call $\bullet_Y$ *comp-convolution* with respect to Y.

3.3.2 Characterization of Elements in $\mathcal{T}(M)$

The convolution operator introduced in the previous subsection helps us characterize/define elements in $\mathcal{T}(M)$, namely, we have the following important property.

Theorem 3.1 $\mathcal{F} \in \mathcal{T}(M)$ *if and only if* $\mathcal{F} * \mathbf{k}_{M \times [0,\infty)} = \mathcal{F}$, *and if and only if* $\mathcal{F} * \mathbf{k}_{M \times (0,\infty)} = 0$.

The proof of this theorem starts with the following observation. From the exact triangle $\mathbf{k}_{[0,\infty)} \to \mathbf{k}_{\{0\}} \to \mathbf{k}_{(0,\infty)}[1] \xrightarrow{+1}$, we get a decomposition

$$\mathcal{F} * \mathbf{k}_{M \times [0,\infty)} \longrightarrow \mathcal{F} \longrightarrow \mathcal{F} * \mathbf{k}_{M \times (0,\infty)}[1] \xrightarrow{+1} \qquad (3.3)$$

Lemma 3.1 $\mathcal{F} * \mathbf{k}_{M \times [0,\infty)} \in \mathcal{T}(M)$.

Proof Take any element in $\mathcal{D}_{\{\tau \leq 0\}}(\mathbf{k}_{M \times \mathbb{R}})$ and assume that $\mathcal{G}$ is a complex of sheaves representing this element. Up to a limit and direct sums, it suffices to consider the open set $U \times (a, b) \subset M \times \mathbb{R}$ (as a basis element) and the sheaf $\mathcal{G} = \mathbf{k}_{U \times (a,b)}$. Then

$$R\mathrm{Hom}(\mathbf{k}_{U \times (a,b)} * \mathbf{k}_{M \times [0,\infty)}, \mathcal{G}) = R\mathrm{Hom}(\mathbf{k}_{U \times [b,\infty)}[-1], \mathcal{G})$$

$$= \mathrm{Cone}(R\Gamma(U \times \mathbb{R}, \mathcal{G}) \to R\Gamma(U \times (-\infty, b), \mathcal{G}))$$

$$= 0 \qquad \text{(by the microlocal Morse lemma)}.$$

Therefore, $\mathcal{F} * \mathbf{k}_{M \times [0,\infty)} \in \mathcal{D}_{\{\tau \leq 0\}}(\mathbf{k}_{M \times \mathbb{R}})^{\perp,1} = \mathcal{T}(M)$.   □

It is easy to check that $SS(\mathcal{F} * \mathbf{k}_{M \times (0,\infty)}[1]) \subset \mathcal{D}_{\{\tau \leq 0\}}(\mathbf{k}_{M \times \mathbb{R}})$ (or see via the geometric meaning of $*$ in Sect. 3.4). Thus (3.3) actually gives an "orthogonal" decomposition.

Remark 3.9 By Proposition 3.1, the orthogonal decomposition (3.3) is equivalent to the fact that the inclusion $\mathcal{D}_{\{\tau \leq 0\}}(\mathbf{k}_{M \times \mathbb{R}}) \hookrightarrow \mathcal{D}(\mathbf{k}_{M \times \mathbb{R}})$ has a left adjoint $p : \mathcal{D}(\mathbf{k}_{M \times \mathbb{R}}) \to \mathcal{D}_{\{\tau \leq 0\}}(\mathbf{k}_{M \times \mathbb{R}})$. Therefore, by Corollary 3.1, for any $\mathcal{F} \in \mathcal{T}(M)$, $p(\mathcal{F}) = 0$. More accurately, p is realized by $*\mathbf{k}_{M \times (0,\infty)}[1]$. Symmetrically, for any $\mathcal{G} \in \mathcal{D}_{\{\tau \leq 0\}}(\mathbf{k}_{M \times \mathbb{R}})$, $\mathcal{G} * \mathbf{k}_{M \times [0,\infty)} = 0$.

Proof (*of Theorem* 3.1) "⇐", by Lemma 3.1. "⇒", by Remark 3.9. □

Corollary 3.2 *The sheaf convolution* $*$ *is well defined in* $\mathcal{T}(M)$.

Proof For any $\mathcal{F}, \mathcal{G} \in \mathcal{T}(M)$,

$$\mathcal{F} * \mathcal{G} = \mathcal{F} * \mathbf{k}_{M \times [0,\infty)} * \mathcal{G} * \mathbf{k}_{M \times [0,\infty)} = \mathcal{F} * \mathcal{G} * \mathbf{k}_{M \times [0,\infty)}.$$

Therefore, $\mathcal{F} * \mathcal{G} \in \mathcal{T}(M)$. □

Remark 3.10 (Remark by F. Zapolsky) The same argument as the proof of Corollary 3.2 above implies that, for any $\mathcal{F} \in \mathcal{T}(M)$, $\mathcal{F} * \mathcal{G} \in \mathcal{T}(M)$ for all $\mathcal{G} \in \mathcal{D}(\mathbf{k}_{M \times \mathbb{R}})$. Therefore, $\mathcal{T}(M)$ is a two-sided ideal in $\mathcal{D}(\mathbf{k}_{M \times \mathbb{R}})$ with respect to sheaf convolution $*$.

3.4 Geometry of Convolution

Whenever we talk about the geometry of an element $\mathcal{F} \in \mathcal{D}(\mathbf{k}_X)$ for some space X, we always mean the behavior of its singular support $SS(\mathcal{F})$ in T^*X. Moreover, as explained/proved in Sect. 2.9, different sheaf operators intertwine with various subset operators on the associated singular supports. By (3.2), $*$ (or $*_{\mathrm{np}}$) is a combination of three operators: $\boxtimes$, $Rs_!$ (or Rs_*) and δ^{-1}. Meanwhile, Corollary 2.1, Example 2.21 and Exercise 2.15 tell us that (i) $\boxtimes$ simply corresponds to product, (ii) the summation map $s : \mathbb{R} \times \mathbb{R} \to \mathbb{R}$ induces the diagonal embedding on co-vector space $s^* : \mathbb{R}^* \to \mathbb{R}^* \times \mathbb{R}^*$, that is,

$$s^*(\tau) = (\tau, \tau),$$

and (iii) the diagonal embedding $\delta : M \to M \times M$ induces the summation on co-vector space $\delta^* : \mathbb{R}^n \times \mathbb{R}^n \to \mathbb{R}^n$,

$$\delta^*(\xi_1, \xi_2) = \xi_1 + \xi_2.$$

Once we combine all these three operators, the following definition is motivated.

Definition 3.6 Let $X, Y \subset T^*(M \times \mathbb{R})$ be two subsets. Denote by $\pi_M : T^*(M \times \mathbb{R}) \to M$ the projection onto the M-component, and similarly for $\pi_{\mathbb{R}^*} : T^*(M \times \mathbb{R}) \to \mathbb{R}^*$ where $\mathbb{R}^*$ stands for the set of co-vectors of the $\mathbb{R}$-component. Define the *set convolution operator* $\hat{*}$ by

$$X \hat{*} Y = \left\{ (m, \xi, t, \tau) \, \middle| \, \begin{array}{c} m \in \pi_M(X) \cap \pi_M(Y) \\ \tau \in \pi_{\mathbb{R}^*}(X) \cap \pi_{\mathbb{R}^*}(Y) \\ \xi = \xi_X + \xi_Y \\ t = t_X + t_Y \end{array} \right\},$$

where $\xi = \xi_X + \xi_Y$ means that over the common point $m \in \pi_M(X) \cap \pi_M(Y)$, one takes the sum of two co-vectors of the M-component from X and Y, and similarly for $t = t_X + t_Y$.

The following result is Proposition 3.13 in [27], and it illustrates the "geometry" of sheaf convolution.

Proposition 3.2 *For $\mathcal{F}, \mathcal{G} \in \mathcal{D}(\mathbf{k}_{M \times \mathbb{R}})$,*

$$SS(\mathcal{F} * \mathcal{G}) \subset SS(\mathcal{F}) \mathbin{\hat{*}} SS(\mathcal{G}). \tag{3.4}$$

Example 3.8 Let $M = \{\text{pt}\}$, $X = \{(0, \tau) \mid \tau \leq 0\} \cup \{(t, 0) \mid 0 \leq t \leq 1\} \cup \{(1, \tau) \mid \tau \geq 0\}$ and $Y = \{(0, \tau) \mid \tau \geq 0\} \cup \{(t, 0) \mid t \geq 0\}$. Then, by the above definition of the set convolution operator,

$$X \mathbin{\hat{*}} Y = \{(1, \tau) \mid \tau \geq 0\} \cup \{(t, 0) \mid t \geq 0\}.$$

For visualization, see Figure 3.1. In fact, X and Y can be realized as singular supports of sheaves: $X = SS(\mathbf{k}_{(0,1)})$ and $Y = SS(\mathbf{k}_{[0,\infty)})$. Then, by Exercise 3.1, $\mathbf{k}_{(0,1)} * \mathbf{k}_{[0,\infty)} = \mathbf{k}_{[1,\infty)}[-1]$, where

$$SS(\mathbf{k}_{[1,\infty)}[-1]) = \{(1, \tau) \mid \tau \geq 0\} \cup \{(t, 0) \mid t \geq 1\} \subsetneqq X \mathbin{\hat{*}} Y.$$

This supports Proposition 3.2. This also shows that in general (3.4) is a strict inclusion.

Remark 3.11 When the summation map s is proper on the supports of the sheaves it acts on, $SS(\mathcal{F} *_{np} \mathcal{G})$ also satisfies the conclusion in Proposition 3.2.

Remark 3.12 It is well-known that geometrically sheaf composition is represented by what is called *Lagrangian correspondence*. This is suggested by the following formula which, under a certain hypothesis (see (1.11) in [26]), says that

$$SS(\mathcal{F} \circ G) \subset SS(\mathcal{F}) \circ SS(\mathcal{G}). \tag{3.5}$$

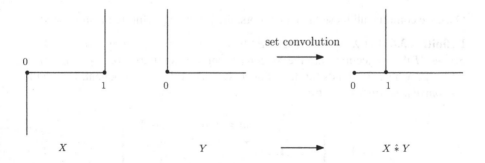

Fig. 3.1 Example of set convolution

The geometric composition operator on the right-hand side of (3.5) is defined as follows. Given two Lagrangian submanifolds $\Lambda_{12} \subset T^*X_1 \times T^*X_2$ and $\Lambda_{23} \subset T^*X_2 \times T^*X_3$, define their *composition* as

$$\Lambda_{13} := \Lambda_{12} \circ \Lambda_{23} := \left\{ ((q_1, p_1), (q_3, p_3)) \;\middle|\; \begin{array}{l} \text{there exists } (q_2, p_2) \in T^*X_2 \text{ s.t.} \\ ((q_1, p_1), (q_2, p_2)) \in \Lambda_{12} \\ ((q_2, -p_2), (q_3, p_3)) \in \Lambda_{23} \end{array} \right\},$$

(3.6)

which is a Lagrangian submanifold in $T^*X_1 \times T^*X_3$. Note that by using the projections $q_{ij} : T^*X_1 \times T^*X_2 \times T^*X_3 \to T^*X_i \times T^*X_j$ (where $1 \le i, j \le 3$), we can rewrite (3.6) as

$$\Lambda_{13} = q_{13}(q_{12}^{-1}(\Lambda_{12}) \cap q_{23}^{-1}(\Lambda_{23})^{a_2})$$

(3.7)

where the upper index a_2 indicates that the co-vectors in the second component are taken with the minus sign.

Exercise 3.2 Check that the comp-convolution operator $\bullet_Y$ (introduced in Definition 3.5) has the following geometric meaning: for any $\mathcal{F} \in \mathcal{D}(\mathbf{k}_{X \times Y \times \mathbb{R}})$ and $\mathcal{G} \in \mathcal{D}(\mathbf{k}_{Y \times Z \times \mathbb{R}})$,

$$SS(\mathcal{F} \bullet_Y \mathcal{G}) \subset \left\{ (x, \xi, z, \theta, t_1 + t_2, \tau) \;\middle|\; \text{there exists some } (y, \eta) \text{ such that } (**) \text{ holds} \right\}$$

where condition $(**)$ is (i) (x, ξ, z, θ) is in the Lagrangian correspondence with respect to $(y, \eta) \in T^*\mathbb{R}_1^n$, and (ii) sum $t_1 + t_2$ in the base is taken over the common point τ.

3.5 Lagrangian Tamarkin Categories

In this section, we provide examples of Tamarkin categories $\mathcal{T}_A(M)$ when the restriction subset A is a Lagrangian submanifold of M. These will be generalizations of the case $A = 0_M$ treated in Example 3.4, and will be the main resources for understanding and testing upcoming abstract propositions and conclusions concerning Tamarkin categories.

3.5.1 $\mathcal{T}_A(M)$ when A is a Lagrangian Submanifold

Example 3.9 (graph case) Let $A = \text{graph}(df)$ for a differentiable function $f : M \to \mathbb{R}$. We will construct a canonical element $\mathcal{F}_f$ in $\mathcal{T}_A(M)$. Here "canonical" means that the reduction $\rho(SS(\mathcal{F}_f)) = \text{graph}(df)$. Note that this is a remarkable step since we have the following useful correspondence:

Fig. 3.2 The picture of the
fiber over (m, t)

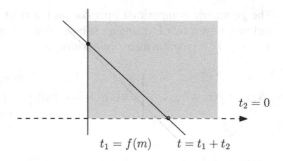

$$t_2 = 0$$

$$t_1 = f(m) \qquad t = t_1 + t_2$$

$$\text{function } f \longleftrightarrow \text{sheaf } \mathcal{F}_f \longleftrightarrow \text{conical subset } SS(\mathcal{F}_f)$$

To construct this $\mathcal{F}_f$, consider the subset

$$N_f = \{(m, t) \subset M \times \mathbb{R} \mid f(m) + t \geq 0\}.$$

We have the following standard proposition (cf. Example 2.16).

Proposition 3.3 *Let ϕ be a smooth function on X and $d\phi(x) \neq 0$ whenever*
$\phi(x) = 0$. Let

$$U = \{x \in X \mid \phi(x) > 0\} \quad and \quad Z = \{x \in X \mid \phi(x) \geq 0\}.$$

Then

- $SS(\mathbf{k}_U) = 0_U \cup \{(x, \tau d\phi(x)) \mid \tau \leq 0, \phi(x) = 0\}$;
- $SS(\mathbf{k}_Z) = 0_Z \cup \{(x, \tau d\phi(x)) \mid \tau \geq 0, \phi(x) = 0\}$.

Applying this proposition to $X = M \times \mathbb{R}$ and $\phi(m, t) = f(m) + t$, if we define
$\mathcal{F}_f := \mathbf{k}_{N_f}$, then

$$SS(\mathcal{F}_f) = \{(m, \tau df(m), -f(m), \tau) \mid \tau \geq 0\} \cup 0_{N_f}. \tag{3.8}$$

Note that the reduction of $SS(\mathcal{F}_f)$ is equal to graph(df). Finally, we need to check
that $\mathcal{F}_f$ is indeed in $\mathcal{T}(M)$, to which purpose we apply Theorem 3.1. Let us compute
$\mathcal{F}_f * \mathbf{k}_{M \times (0, \infty)}$:

$$\mathcal{F}_f * \mathbf{k}_{M \times (0, \infty)} = \mathbf{k}_{N_f} * \mathbf{k}_{M \times (0, \infty)}$$

$$= \delta^{-1} Rs_! \mathbf{k}_{\{(m_1, m_2, t_1, t_2) \mid f(m_1) + t_1 \geq 0, \, t_2 > 0\}}.$$

At each stalk $(m, t) \in M \times \mathbb{R}$, Figure 3.2 shows the computation picture on fibers.
Since it always cuts out a finite half-open and half-closed interval, the compactly
supported cohomology vanishes. Therefore, $\mathcal{F}_f * \mathbf{k}_{M \times (0, \infty)} = 0$.

Example 3.10 (generating function) We can generalize the construction in the previous example when $L \subset T^*M$ admits a generating function, i.e., there exists a function $S : M \times \mathbb{R}^K \to \mathbb{R}$ for some $K \in \mathbb{N}$ such that

$$L = \{(m, \partial_m S(m, \xi)) \mid \partial_\xi S(m, \xi) = 0\}.$$

Consider $N_S = \{(m, \xi, t) \mid S(m, \xi) + t \geq 0\}$. Let $p : M \times \mathbb{R}^K \times \mathbb{R} \to M \times \mathbb{R}$ be the canonical projection (and, for brevity, we do not specify the dependence of p on the dimension K). It has been checked in [61], Subsect. 1.2. on page 112, that $\rho(SS(Rp_!\mathbf{k}_{N_S})) = L$. In other words, $\mathcal{F}_S := Rp_!\mathbf{k}_{N_S}$ is the canonical sheaf associated to this L. For the reader's convenience, we provide a concrete example in this set-up.

Let $M = \mathbb{R}$ and $L = 0_\mathbb{R}$. Then L has a generating function $S : \mathbb{R} \times \mathbb{R} \to \mathbb{R}$ given by

$$S(m, \xi) = \xi^2, \qquad \text{usually called quadratic at infinity.}$$

Indeed, this is a generating function for $0_\mathbb{R}$ since if $\partial_\xi S(m, \xi) = 0$, then $\xi = 0$. So $(m, \partial_m S) = (m, 0)$. By our construction,

$$N_S = \{(m, \xi, t) \in \mathbb{R}^3 \mid \xi^2 + t \geq 0\}.$$

Take $\mathcal{F}_S = Rp_!\mathbf{k}_{N_S}$ where $p : \mathbb{R}^3 \to \mathbb{R}_1 \times \mathbb{R}_3$, and let us describe its stalks. For any (m, t),

$$(\mathcal{F}_S)_{(m,t)} = H_c^*(\mathbb{R}, \mathbf{k}_{N_S}|_{p^{-1}(m,t)}).$$

There are two cases of the fiber $p^{-1}(m, t)$ (see Figure 3.3). In Case 1,

$$H_c^*(\mathbb{R}, \mathbf{k}_{N_S}|_{p^{-1}(m,t)}) = H_c^*(\mathbb{R}, \mathbf{k}_\mathbb{R}) = \mathbf{k}[-1].$$

In Case 2,

$$H_c^*(\mathbb{R}, \mathbf{k}_{N_S}|_{p^{-1}(m,t)}) = H_c^*(\mathbb{R}, \mathbf{k}_{(-\infty, -\sqrt{t}]} \oplus \mathbf{k}_{[\sqrt{t}, \infty)})$$
$$= H_c^*(\mathbb{R}, \mathbf{k}_{(-\infty, -\sqrt{t}]}) \oplus H_c^*(\mathbb{R}, \mathbf{k}_{[\sqrt{t}, \infty)}) = 0.$$

In fact, one can check that $\mathcal{F}_S = \mathbf{k}_{\{(m,t) \mid t \geq 0\}} = \mathbf{k}_{\mathbb{R} \times [0,\infty)}$, and its reduction is just $0_\mathbb{R}$.

The next example motives Separation Theorem which will be explained later (see Theorem 3.3 in Section 3.7).

Example 3.11 (Disjoint reductions) Let $M = \mathbb{R}$ and $A = 0_\mathbb{R}$, $B = \mathbb{R} \times \{1\} \subset T^*\mathbb{R}$. Note that $A \cap B = \emptyset$. Since both A and B can be realized as graphs of differentials, more specifically, $A = \text{graph}(df)$ with $f \equiv 0$ and $B = \text{graph}(dg)$ where $g(m) = m$, we can pass to the relation between their canonical sheaves. By our construction,

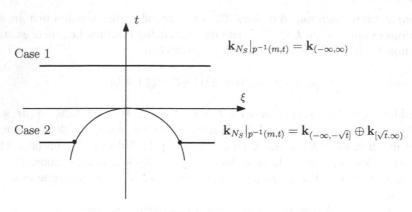

Fig. 3.3 Two different cases of the fiber over (m, t)

$\mathcal{F}_f = \mathbf{k}_{M \times [0,\infty)} \in \mathcal{T}_A(M)$ and $\mathcal{F}_g = \mathbf{k}_{\{(m,t) \mid m+t \geq 0\}} \in \mathcal{T}_B(M)$. It can be checked that $R\mathrm{Hom}(\mathcal{F}_f, \mathcal{F}_g) = 0$. Since we are working over a field $\mathbf{k}$, it is a standard fact (see Definition 13.1.18 and Example 13.1.19 in [33]) that at most two terms from $R\mathrm{Hom}(\mathcal{F}_f, \mathcal{F}_g)$ are non-trivial, namely, $R^0\mathrm{Hom}(\mathcal{F}_f, \mathcal{F}_g) = \mathrm{Hom}(\mathcal{F}_f, \mathcal{F}_g)$ and $R^1\mathrm{Hom}(\mathcal{F}_f, \mathcal{F}_g)$ (usually denoted as $\mathrm{Ext}^1(\mathcal{F}_f, \mathcal{F}_g)$). That $R^0\mathrm{Hom}(\mathcal{F}_f, \mathcal{F}_g) = \mathrm{Hom}(\mathcal{F}_f, \mathcal{F}_g) = 0$ is easily seen from the following handy result; the Ext^1 term is harder to check.

Exercise 3.3 Suppose I and J are locally closed subsets of $\mathbb{R}^n$, then

$$\mathrm{Hom}(\mathbf{k}_I, \mathbf{k}_J) = \begin{cases} \mathbf{k}, & \text{if } I \cap J \neq \emptyset \text{ and closed in } I \text{ open in } J, \\ 0, & \text{otherwise.} \end{cases}$$

In particular, if I and J are closed, $\mathrm{Hom}(\mathbf{k}_I, \mathbf{k}_J)$ is non-trivial if and only if $J \subset I$.

Remark 3.13 (Warning!) Exercise 3.3 could be misleading. It shows that $\mathrm{Hom}(\mathbf{k}_{[0,\infty)}, \mathbf{k}_{(-\infty,0)}) = 0$, which gives the impression that no information may be extracted from $\mathbf{k}_{[0,\infty)}$ and $\mathbf{k}_{(-\infty,0)}$. However, $R\mathrm{Hom}(\mathbf{k}_I, \mathbf{k}_J)$ actually contains more information than $\mathrm{Hom}(\mathbf{k}_I, \mathbf{k}_J)$, coming exactly from some higher i-th derived functors. In fact, $R\mathrm{Hom}(\mathbf{k}_{[0,\infty)}, \mathbf{k}_{(-\infty,0)}) = \mathbf{k}[-1]$, non-trivial at degree 1. In general, we have the following result (see Appendix A.2),

Theorem 3.2 *Let $a < b$ and $c < d$ in $\mathbb{R}$. Then*

$$R\mathrm{Hom}(\mathbf{k}_{[a,b)}, \mathbf{k}_{[c,d)}) = \begin{cases} \mathbf{k}, & \text{for } a \leq c < b \leq d, \\ \mathbf{k}[-1], & \text{for } c < a \leq d < b, \\ 0, & \text{otherwise.} \end{cases}$$

Example 3.12 $R\mathrm{Hom}(\mathbf{k}_{[a,b)}, \mathbf{k}_{[T,\infty)}) = \mathbf{k}$ when $T \in [a, b)$, and zero otherwise.

Example 3.13 $R\mathrm{Hom}(\mathbf{k}_{[0,\infty)}, \mathbf{k}_{[c,d)}) = \mathbf{k}[-1]$ when $c < 0 \leq d < +\infty$, and zero otherwise. Here c can be $-\infty$. When $d = \infty$, $R\mathrm{Hom}(\mathbf{k}_{[0,\infty)}, \mathbf{k}_{[c,\infty)}) = \mathbf{k}$ when $c \geq 0$, and zero otherwise.

3.5.2 Convolution and Lagrangians

The next example reveals the geometric meaning of sheaf convolution $*$ in term of Lagrangians.

Example 3.14 Let $f, g : M \to \mathbb{R}$ be two smooth functions. Then

$$SS(\mathcal{F}_f) = \{(m, \tau df(m), -f(m), \tau) \mid \tau \geq 0\} \cup 0_{N_f}$$

and

$$SS(\mathcal{F}_g) = \{(m, \tau dg(m), -g(m), \tau) \mid \tau \geq 0\} \cup 0_{N_g}.$$

By Proposition 3.2,

$$SS(\mathcal{F}_f * \mathcal{F}_g) \subset SS(\mathcal{F}_f) \hat{*} SS(\mathcal{F}_g)$$

$$= \{(m, \tau(df(m)+dg(m)), -f(m)-g(m), \tau)|\tau{\geq}0\}\cup\{\text{some 0-section}\}$$

$$= \{(m, \tau d(f + g)(m), -(f + g)(m), \tau) \mid \tau \geq 0\} \cup \{\text{some 0-section}\}.$$

Then the reduction $\rho(SS(\mathcal{F}_f) \hat{*} SS(\mathcal{F}_g)) = \mathrm{graph}(d(f + g))$. This can be regarded as the *fiberwise summation* of $\mathrm{graph}(df)$ and $\mathrm{graph}(dg)$. In fact, it can checked that $\mathcal{F}_f * \mathcal{F}_g = \mathcal{F}_{f+g}$ (see the proof of Proposition 3.4 below).

In the spirit of Example 3.10, we can also consider examples involving generating functions. Let L_1, L_2 be two Lagrangians submanifolds of T^*M admitting generating functions $S_1 : M \times \mathbb{R}^{\ell_1} \to \mathbb{R}$ and $S_2 : M \times \mathbb{R}^{\ell_2} \to \mathbb{R}$, respectively. Recall that $\mathcal{F}_{S_1}$ and $\mathcal{F}_{S_2}$ denote the associated canonical sheaves constructed in Example 3.10. Then we have the following result.

Proposition 3.4 *The convolution $\mathcal{F}_{S_1} * \mathcal{F}_{S_2}$ is a canonical sheaf of the fiberwise sum of L_1 and L_2, denoted by $L_1 +_b L_2$.*

In order to prove Proposition 3.4, we introduce the following "fiberwise summation" generating function $S : M \times \mathbb{R}^{\ell_1+\ell_2} \to \mathbb{R}$ by the formula

$$S(m, \xi_1, \xi_2) = S_1(m, \xi_1) + S_2(m, \xi_2). \tag{3.9}$$

Observe that this S is a generating function of $L_1 +_b L_2$. In fact, for $\xi = (\xi_1, \xi_2)$, the condition $\partial_\xi S(x, \xi) = 0$ is equivalent to $\partial_{\xi_1} S(x, \xi) = \partial_{\xi_2} S(x, \xi) = 0$. Then for those (x, ξ) we have $\partial_x S(x, \xi) = \partial_x S_1(x, \xi_1) + \partial_x S_2(x, \xi_2)$, and this is exactly the summation of co-vectors of L_1 and L_2. Similarly, we can consider the canonical sheaf $\mathcal{F}_S$ of S constructed in Example 3.9 and the projection map $p : M \times \mathbb{R}^{\ell_1+\ell_2} \times \mathbb{R} \to M \times \mathbb{R}$.

Proof *(of Proposition 3.4)* We will show that $\mathcal{F}_{S_1} * \mathcal{F}_{S_2} = \mathcal{F}_S$, which yields the desired conclusion. Note that the projection $p' : M \times M \times \mathbb{R}^{\ell_1+\ell_2} \times \mathbb{R} \times \mathbb{R} \to M \times M \times \mathbb{R} \times \mathbb{R}$ can be regarded as the product $p_1 \times p_2$, where $p_i : M \times \mathbb{R}^{\ell_i} \times \mathbb{R} \to M \times \mathbb{R}$ for $i = 1, 2$. Then, the diagram

$$
\begin{array}{ccccc}
M \times \mathbb{R}^{\ell_1} \times \mathbb{R} & \xleftarrow{\pi_1} & M \times M \times \mathbb{R}^{\ell_1+\ell_2} \times \mathbb{R} \times \mathbb{R} & \xrightarrow{\pi_2} & M \times \mathbb{R}^{\ell_2} \times \mathbb{R} \\
\downarrow{\scriptstyle p_1} & & \downarrow{\scriptstyle p'} & & \downarrow{\scriptstyle p_2} \\
M \times \mathbb{R} & \xleftarrow{\tilde{\pi}_1} & M \times M \times \mathbb{R} \times \mathbb{R} & \xrightarrow{\tilde{\pi}_2} & M \times \mathbb{R}
\end{array}
$$

and Exercise II.18 (i) in [32] show that

$$(\tilde{\pi}_1^{-1} Rp_{1!}\mathbf{k}_{\{S_1+t_1 \geq 0\}}) \otimes (\tilde{\pi}_2^{-1} Rp_{2!}\mathbf{k}_{\{S_2+t_2 \geq 0\}}) = Rp'_!(\pi_1^{-1}\mathbf{k}_{\{S_1+t_1 \geq 0\}} \otimes \pi_2^{-1}\mathbf{k}_{\{S_2+t_2 \geq 0\}}).$$

Next, the commutative diagram,

$$
\begin{array}{ccccc}
M \times \mathbb{R}^{\ell_1+\ell_2} \times \mathbb{R} & \xrightarrow{\delta} & M \times M \times \mathbb{R}^{\ell_1+\ell_2} \times \mathbb{R} & \xleftarrow{s} & M \times M \times \mathbb{R}^{\ell_1+\ell_2} \times \mathbb{R} \times \mathbb{R} \\
\downarrow{\scriptstyle p} & & \downarrow{\scriptstyle p''} & & \downarrow{\scriptstyle p'} \\
M \times \mathbb{R} & \xrightarrow{\delta} & M \times M \times \mathbb{R} & \xleftarrow{s} & M \times M \times \mathbb{R} \times \mathbb{R}
\end{array}
$$

shows that

$$
\begin{aligned}
\mathcal{F}_{S_1} * \mathcal{F}_{S_2} &= \delta^{-1} Rs_!(\tilde{\pi}_1^{-1} Rp_{1!}\mathbf{k}_{\{S_1+t_1 \geq 0\}}) \otimes (\tilde{\pi}_2^{-1} Rp_{2!}\mathbf{k}_{\{S_2+t_2 \geq 0\}}) \\
&= \delta^{-1} Rs_! Rp'_!(\pi_1^{-1}\mathbf{k}_{\{S_1+t_1 \geq 0\}} \otimes \pi_2^{-1}\mathbf{k}_{\{S_2+t_2 \geq 0\}}) \\
&= \delta^{-1} Rp''_! Rs_!(\pi_1^{-1}\mathbf{k}_{\{S_1+t_1 \geq 0\}} \otimes \pi_2^{-1}\mathbf{k}_{\{S_2+t_2 \geq 0\}}) \\
&= Rp_! \delta^{-1} Rs_!(\pi_1^{-1}\mathbf{k}_{\{S_1+t_1 \geq 0\}} \otimes \pi_2^{-1}\mathbf{k}_{\{S_2+t_2 \geq 0\}}) \\
&= Rp_!\mathbf{k}_{\{S+t \geq 0\}} = \mathcal{F}_S.
\end{aligned}
$$

The third equality comes from the commutativity of the right square in the diagram above. The fourth equality comes from the base change formula from the left square in the diagram above. The fifth equality is a direct computation. $\square$

3.6   Shift Functor and Torsion Elements

One of the advantages of the extra variable $\mathbb{R}$ in Tamarkin categories is that this $\mathbb{R}$-component provides a 1-dimensional ruler for filtrations, and then objects can be shifted. For a multi-dimensional ruler, see a parallel theory developed in [27].

Definition 3.7 For any $a \in \mathbb{R}$, consider the map $T_a : M \times \mathbb{R} \to M \times \mathbb{R}$ defined by $T_a(m, t) = (m, t + a)$. It induces a map on $\mathcal{D}(\mathbf{k}_{M \times \mathbb{R}})$, denoted by T_{a*}. Note that, on the level of stalks, $(T_{a*}\mathcal{F})_{(m,t)} = \mathcal{F}_{(m,t-a)}$.

Exercise 3.4 For any $\mathcal{F} \in \mathcal{D}(\mathbf{k}_{M \times \mathbb{R}})$, $T_{a*}\mathcal{F} = \mathcal{F} * \mathbf{k}_{M \times \{a\}}$.

Lemma 3.2 (1) T_{a*} is well-defined over $\mathcal{T}(M)$. (2) For any $a \leq b$, there exists a natural transformation $\tau_{a,b} : T_{a*} \to T_{b*}$. In particular, for any $c \geq 0$, there exists a natural transformation $\tau_c : \mathbb{1} \to T_{c*}$.

Proof (1) For any $\mathcal{F} \in \mathcal{T}(M)$, we have

$$T_{a*}\mathcal{F} * \mathbf{k}_{M \times [0,\infty)} = \mathcal{F} * \mathbf{k}_{M \times \{a\}} * \mathbf{k}_{M \times [0,\infty)}$$

$$= \mathcal{F} * \mathbf{k}_{M \times [0,\infty)} * \mathbf{k}_{M \times \{a\}}$$

$$= \mathcal{F} * \mathbf{k}_{M \times \{a\}} = T_{a*}\mathcal{F}.$$

In particular, for any $\mathcal{F} \in \mathcal{T}(M)$, $T_{a*}\mathcal{F} = \mathcal{F} * \mathbf{k}_{M \times [a,\infty)}$. (2) The natural transformation $\tau_{a,b}$ is given by the restriction $\mathbf{k}_{M \times [a,\infty)} \to \mathbf{k}_{M \times [b,\infty)}$. $\qquad\square$

Definition 3.8 We call $\mathcal{F} \in \mathcal{T}(M)$ a *c-torsion* element if $\tau_c(\mathcal{F}) : \mathcal{F} \to T_{c*}\mathcal{F}$ is zero.

Remark 3.14 Note that the torsion elements can be defined in a bigger category $\mathcal{D}_{\{\tau \geq 0\}}(M \times \mathbb{R})$. Explicitly, Proposition 4.1 in [1] or Proposition 5.2.3 (together with Proposition 3.5.4) in [32] say that $\mathcal{F} \in \mathcal{D}_{\{\tau \geq 0\}}(M \times \mathbb{R})$ if and only if $\mathcal{F} *_{\mathrm{np}} \mathbf{k}_{M \times [0,\infty)}$ is isomorphic to $\mathcal{F}$. Then again we have a well-defined map (still denoted by) $\tau_c(\mathcal{F})$ which is induced by the restriction map $\mathbf{k}_{[0,\infty)} \to \mathbf{k}_{[c,\infty)}$.

Example 3.15 Here are some examples of torsion and non-torsion elements.

(1) When $M = \{\mathrm{pt}\}$, $\mathbf{k}_{[a,b)}$ with $b < \infty$ is a $(b - a)$-torsion.
(2) For $\mathbf{k}_{M \times [0,\infty)} \in \mathcal{T}_{0_M} M$, $\mathcal{F}$ is non-torsion because

$$\tau_c(\mathcal{F}) : \mathbf{k}_{M \times [0,\infty)} \to T_{c*}(\mathbf{k}_{M \times [0,\infty)})(= \mathbf{k}_{M \times [c,\infty)})$$

is a non-trivial morphism for any $c \geq 0$.

The following example shows one case in symplectic topology that is difficult to study by classical tools, but can be easily handled in the language of sheaves. According to the argument in [1], it will be essential to the proof of the positivity of displacement energies (see Sect. 4.3).

Fig. 3.4 Whitney immersion

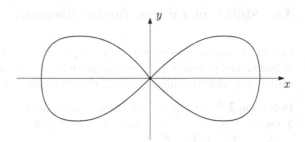

Fig. 3.5 Front projection and a torsion sheaf

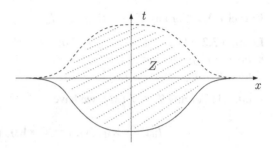

Example 3.16 (Lagrangian torsion in [1]) Let $M = \mathbb{R}$ and consider the following immersed Lagrangian submanifold L in $T^*\mathbb{R}$. It can be generalized to higher dimensional space, and is generally called a *Whitney immersion* (see Figure 3.4). Consider an intersection of two domains as in Figure 3.5. Denote the shaded region as Z. Then, by Sect. 3.5, we know that $\rho(SS(\mathbf{k}_Z)) = L$ simply by tracking slopes of the boundary pointwisely. More importantly, $\mathbf{k}_Z$ is a torsion element since its support is a bounded region and a sufficiently large shift along the t-direction implies that $\tau_c(\mathbf{k}_Z) = 0$.

Remark 3.15 Since every Darboux ball B contains a rescaled L as the one in Figure 3.4, for $\mathcal{T}_B(M)$, there always exist elements in $\mathcal{T}_B(M)$ as Example 3.16. On the other hand, if the shift c is not big enough, say $c < 2\max\{t \mid (x, t) \in Z\}$, then the map

$$\tau_c(\mathbf{k}_Z) : \mathbf{k}_Z \to \mathbf{k}_{Z+c} \quad \text{where } Z + c \text{ is a shift in the } t\text{-direction.}$$

is non-trivial by Exercise 3.3.

3.7   Separation Theorem and the Adjoint Sheaf

3.7.1   Restatement of the Separation Theorem

A fundamental result in Tamarkin category is the following Separation Theorem, which generalizes the obvious fact that $\mathrm{Hom}(\mathcal{F}, \mathcal{G}) = 0$ if $\mathrm{supp}(\mathcal{F}) \cap \mathrm{supp}(\mathcal{G}) = \emptyset$.

Fig. 3.6 Non-trivial case for the Separation Theorem

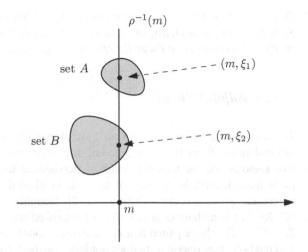

Theorem 3.3 *Let A, B be two compact subsets in T^*M. If $A \cap B = \emptyset$, then for any $\mathcal{F} \in \mathcal{T}_A(M)$ and $\mathcal{G} \in \mathcal{T}_B(M)$, $\mathrm{RHom}(\mathcal{F}, \mathcal{G}) = 0$.*

Remark 3.16 This theorem is harder to prove than it may seem, because the supports of $\mathcal{F}$ and $\mathcal{G}$ actually lie in $\overline{\rho^{-1}(A)}$ and $\overline{\rho^{-1}(B)}$, respectively. Note that $A \cap B = \emptyset$ does *not* guarantee that $\overline{\rho^{-1}(A)} \cap \overline{\rho^{-1}(B)} = \emptyset$. In fact, there are two cases if we make this explicit. One, if there is no common $m \in M$ for A and B, then indeed $\overline{\rho^{-1}(A)} \cap \overline{\rho^{-1}(B)} = \emptyset$, and in this case, the Separation Theorem holds trivially. The other case is that A and B do have some common $m \in M$ with co-linear co-vectors, that is, there exist $(m, \xi_1) \in A$ and $(m, \xi_2) \in B$ such that $\xi_2 = \lambda \xi_1$ for some $\lambda > 0$. Then the entire (non-negative) fiber is contained both in $\overline{\rho^{-1}(A)}$ and $\overline{\rho^{-1}(B)}$ by the definition of ρ. Figure 3.6 is a general picture of this case. For this case, a serious proof is needed.

Instead of computing $R\mathrm{Hom}(\mathcal{F}, \mathcal{G})$ directly, we will rewrite it in terms of a new sheaf (people usually call it *internal hom in $\mathcal{T}(M)$*), denoted by $\mathcal{H}om^*(\mathcal{F}, \mathcal{G})$. Here is a precise statement.

Lemma 3.3 *For any $\mathcal{F}, \mathcal{G} \in \mathcal{T}(M)$,*

$$R\mathrm{Hom}(\mathcal{F}, \mathcal{G}) = R\mathrm{Hom}(\mathbf{k}_{[0,\infty)}, R\pi_* \mathcal{H}om^*(\mathcal{F}, \mathcal{G}))) \qquad (3.10)$$

where $\pi : M \times \mathbb{R} \to \mathbb{R}$ is the projection. The sheaf of $\mathcal{H}om^*(\mathcal{F}, \mathcal{G})$ is introduced in Definition 3.10.

Note that $R\pi_* \mathcal{H}om^*(\mathcal{F}, \mathcal{G})$ on the right-hand side of the identity (3.10) is in general a complex of sheaves over $\mathbb{R}$, which reduces considerably the difficulty of the discussion and computations. Moreover, under a constructibility assumption, $R\pi_* \mathcal{H}om^*(\mathcal{F}, \mathcal{G}) = \bigoplus \mathbf{k}_{[a_i, b_i)}[n_i]$ where the type of intervals is determined by the singular support of $\mathcal{H}om^*$. Then we are able to compute the right-hand side of

(3.10) based on Theorem 3.2 or Example 3.13. This will be particularly helpful in Sect. 3.9. Last but not least, we want to emphasize that what the Separation Theorem really establishes is that sheaf $R\pi_* \mathcal{H}om^*(\mathcal{F}, \mathcal{G}) = 0$.

3.7.2 Adjoint Sheaf

To motivate the definition of $\mathcal{H}om^*$, we start with an explanation of the computational result from Example 3.6, i.e., the similarity between the convolution of two sheaves and the tensor product of persistence **k**-modules. One can view any persistence **k**-module in terms of an abstract filtered complex $C_\bullet = (C, \partial_C, \ell_C)$, where ℓ_C is the associated filtration [58]. Roughly speaking, a filtered complex (C, ∂_C, ℓ_C) is a chain complex (C, ∂_C) endowed with a function (called filtration) $\ell_C : C \to \mathbb{R} \cup \{-\infty\}$ such that ℓ_C satisfies a non-Archimedean triangle inequality and the boundary operator ∂_C does not increase the filtration. This is the right chain-level analog of a persistence **k**-module. In many cases, a filtered complex is the one step before taking homologies (to obtain a persistence **k**-module), and in general it contains more information and is easier to work with than persistence **k**-modules. Operations on chain complexes in the standard homological algebra work on filtered chain complexes as well. The following table shows some analogy between filtered chain complexes and sheaves.

sheaves	filtered complex
$\mathcal{F} * \mathcal{G}$	$C_\bullet \otimes D_\bullet$
$\mathcal{H}om^*(\mathcal{F}, \mathcal{G}) :\approx \overline{\mathcal{F}} * \mathcal{G}$	$\mathrm{Hom}(C_\bullet, D_\bullet) \simeq C_\bullet^* \otimes D_\bullet$

where the second row and the third row represent adjoint relations. Any proposed adjoint functor of convolution $*$ requires a well-defined "dual" sheaf $\overline{\mathcal{F}}$ (here "−" on the top is used for our dual here, since there already exists the term dual sheaf, say (ii) in Definition 3.1.16 in [32]). To avoid confusion, we call $\overline{\mathcal{F}}$ the *adjoint sheaf* of $\mathcal{F}$.

Recall that the filtered *dual* complex $C_\bullet^*$ (of the filtered complex $C_\bullet$) is defined as follows:

$$x \xrightarrow{\partial_C} \partial_C x \iff y \xrightarrow{\partial_C^*} \partial_C^* y$$

where y is the dual of $\partial_C x$. The filtration is given by $\ell_{C^*}(y) = -\ell_C(\partial x)$ and $\ell_{C^*}(\partial y) = -\ell_C(x)$. In terms of sheaves, we have

$$\mathbf{k}_{[a,b)} \to \mathbf{k}_{(-b,-a]} = i^{-1}\mathbf{k}_{[a,b)}, \text{ with } i : \mathbb{R} \to \mathbb{R} \text{ defined by } i(t) = -t.$$

However, note that $\mathbf{k}_{(-b,-a]}$ is not an element in $\mathcal{T}(\mathrm{pt})$ (because its singular support belongs to the wrong half-plane). We can use the following proposition (**Exercise**) to correct the open-closed relation at the endpoints,

Proposition 3.5 $R\mathcal{H}om(\mathbf{k}_{[a,b)}, \mathbf{k}_\mathbb{R}) = \mathbf{k}_{(a,b]}$ and $R\mathcal{H}om(\mathbf{k}_{(a,b]}, \mathbf{k}_\mathbb{R}) = \mathbf{k}_{[a,b)}$.

Then we propose the following definition of the adjoint sheaf.

Definition 3.9 Define the adjoint sheaf of $\mathcal{F}$ by,

$$\overline{\mathcal{F}} = R\mathcal{H}om(i^{-1}\mathcal{F}, \mathbf{k}_{M \times \mathbb{R}})[1].$$

Here, degree shift [1] corresponds to the dimension of $\mathbb{R}$-factor in total space $M \times \mathbb{R}$. In other words, if we consider space $M \times \mathbb{R}^m$ (to define Tamarkin categories), then we need to modify our definition of $\mathcal{H}om^*$ to have degree shift $[m]$.

Then our proposed adjoint functor of $*$ is defined as follows.

Definition 3.10 For any $\mathcal{F}, \mathcal{G} \in \mathcal{D}(\mathbf{k}_{M \times \mathbb{R}})$, define

$$\mathcal{H}om^*(\mathcal{F}, \mathcal{G}) := \overline{\mathcal{F}} *_{np} \mathcal{G}$$

$$= \delta^{-1} Rs_*(q_2^{-1} R\mathcal{H}om(i^{-1}\mathcal{F}, \mathbf{k}_{M \times \mathbb{R}}) \otimes q_1^{-1}\mathcal{G})[1].$$

Exercise 3.5 Check that if both $\mathcal{F}, \mathcal{G} \in \mathcal{D}_{\{\tau \geq 0\}}(M \times \mathbb{R})$, then $\mathcal{H}om^*(\mathcal{F}, \mathcal{G}) \in \mathcal{D}_{\{\tau \geq 0\}}(M \times \mathbb{R})$. See Corollary 3.4 for a stronger statement.

Example 3.17 Let $M = \{\text{pt}\}$; then

$$\mathcal{H}om^*(\mathbf{k}_{[a,b)}, \mathbf{k}_{[c,d)}) = \mathbf{k}_{[c-b, \min\{d-b, c-a\})}[1] \oplus \mathbf{k}_{[\max\{d-b, c-a\}, d-a)}.$$

Example 3.18 $\mathcal{H}om^*(\mathbf{k}_{M \times [0, \infty)}, \mathbf{k}_{M \times [0, \infty)}) = \mathbf{k}_{M \times (-\infty, 0)}[1]$. For a general result like this, see (59) in [27].

Example 3.19 (Geometry of $\mathcal{H}om^*$) Let $f, g : M^n \to \mathbb{R}$ be two differentiable functions. Consider the canonical sheaves associated to $\text{graph}(df)$ and $\text{graph}(dg)$, that is, $\mathcal{F} = \mathbf{k}_{\{(m,t) \mid f(m)+t \geq 0\}}$ and $\mathcal{G} = \mathbf{k}_{\{(m,t) \mid g(m)+t \geq 0\}}$, respectively. First, note that

$$\overline{\mathcal{F}} = \mathbf{k}_{\{(m,t) \mid f(m)-t > 0\}}[1].$$

Then, by Proposition 3.3 (open domain case),

$$SS(\overline{\mathcal{F}}) = \{(m, \tau df(m), f(m), -\tau) \mid \tau \leq 0\} \cup 0_{\{(m,t) \mid f(m)-t > 0\}}$$

$$= \{(m, \tau d(-f)(m), -(-f)(m), \tau) \mid \tau \geq 0\} \cup 0_{\{(m,t) \mid f(m)-t > 0\}},$$

whose reduction is simply equal to $\text{graph}(d(-f))$.[1] In other words, the geometric meaning of the adjoint sheaf $\overline{\mathcal{F}}$ is just taking the minus sign fiberwise in the Lagrangian manifold.

---

[1] Note that the canonical sheaf for $\text{graph}(d(-f))$, i.e., $\mathbf{k}_{\{-f(m)+t \geq 0\}}$, does not have the same singular support as $\overline{\mathcal{F}}$. Luckily they only differ in the 0-section part.

Fig. 3.7 Computation of $\mathcal{H}om^*(\mathcal{F}, \mathcal{G})$

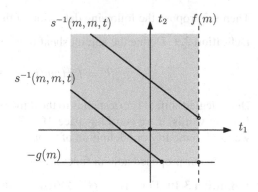

Second, by definition,

$$\mathcal{H}om^*(\mathcal{F}, \mathcal{G}) = \delta^{-1} R s_* \mathbf{k}_{\{(m_1, m_2, t_1, t_2) \mid f(m_1) - t_1 > 0, g(m_2) + t_2 \geq 0\}}[1] := \delta^{-1} R s_* \mathbf{k}_N[1].$$

On each stalk (m, t), Figure 3.7 provides a computational picture for $(\delta^{-1} R s_* \mathbf{k}_N)_{(m,t)} = (R s_* \mathbf{k}_N)_{(m,m,t)} = H^*(\mathbb{R}; \mathbf{k}_N|_{s^{-1}(m,m,t)})$. Therefore, whenever $t < f(m) - g(m)$, we have $(\delta^{-1} R s_* \mathbf{k}_N)_{(m,t)} = \mathbf{k}$ (at degree 0) and zero otherwise. Hence,

$$\mathcal{H}om^*(\mathcal{F}, \mathcal{G}) = \mathcal{H}om^*(\mathcal{F}_f, \mathcal{F}_g) = \mathbf{k}_{\{(m,t) \in M \times \mathbb{R} \mid f(m) - g(m) > t\}}[1]. \qquad (3.11)$$

Roughly speaking, the geometric meaning of $\mathcal{H}om^*$ is the fiberwise subtraction of Lagrangian manifolds. This is actually confirmed by Proposition 3.13 in [27], that is,

$$SS(\mathcal{H}om^*(\mathcal{F}, \mathcal{G})) \subset SS(\mathcal{F}) \overset{*}{*} SS(G)^{\mathrm{a}}$$

where $-^{\mathrm{a}}$ means changing the sign of the corresponding co-vectors.

Remark 3.17 Here is a comparison between Definition 3.10 of $\mathcal{H}om^*$ and the usual definition in the literature, see for instance (3.5) in [1]:

$$\mathcal{H}om^*(\mathcal{F}, \mathcal{G})_{AI} := R s_* R\mathcal{H}om(q_2^{-1} i^{-1} \mathcal{F}, q_1^! \mathcal{G}), \qquad (3.12)$$

where q_i is the projection from $M \times \mathbb{R}_1 \times \mathbb{R}_2$ to $M \times \mathbb{R}_i$ (with the dimension of the fiber equal to 1), and $q_1^!$ is the right adjoint of $Rq_{1!}$ (see Sect. 3.1 in [32]). We claim that $\mathcal{H}om^*(\mathcal{F}, \mathcal{G})_{AI} = \mathcal{H}om^*(\mathcal{F}, \mathcal{G})$. In fact, for the notation in [32], s is just our $\delta^{-1} s$. Then

$$(q_2^{-1} R\mathcal{H}om(i^{-1} \mathcal{F}, \mathbf{k}_{M \times \mathbb{R}}) \otimes q_1^{-1} \mathcal{G})[1] = D(i^{-1} \mathcal{F}) \boxtimes \mathcal{G},$$

where $D(-)$ is the dual sheaf defined by using $R\mathcal{H}om$ (see (ii) in Definition 3.1.16 in [32]). By Proposition 3.4.4 in [32], $D(i^{-1}\mathcal{F}) \boxtimes \mathcal{G} = R\mathcal{H}om(q_2^{-1}i^{-1}\mathcal{F}, q_1^!\mathcal{G})$, which yields the needed identification.

3.8 Proof of the Separation Theorem

The first step is to prove identity (3.10) which needs to confirm that $\mathcal{H}om^*$ is indeed the right adjoint of the operator $*$. Explicitly,

Proposition 3.6 *For any $\mathcal{F}_1, \mathcal{F}_2$ and $\mathcal{F}_3$ in $\mathcal{D}(\mathbf{k}_{M \times \mathbb{R}})$, we have*

$$R\text{Hom}(\mathcal{F}_1 * \mathcal{F}_2, \mathcal{F}_3) = R\text{Hom}(\mathcal{F}_1, \mathcal{H}om^*(\mathcal{F}_2, \mathcal{F}_3)).$$

This has been proved essentially by Lemma 3.10 in [27]. It provides a third expression of $\mathcal{H}om^*(\mathcal{F}, \mathcal{G})$ which is equivalent to both Definition 3.10 and (3.12). Here we give an elementary example in the support of this proposition.

Example 3.20 Let $\mathcal{F}_1 = \mathcal{F}_2 = \mathcal{F}_3 = \mathbf{k}_{[0,\infty)}$. Then we know that $\mathcal{F}_1 * \mathcal{F}_2 = \mathbf{k}_{[0,\infty)}$ and $\mathcal{H}om^*(\mathcal{F}_2, \mathcal{F}_3) = \mathbf{k}_{(-\infty,0)}[1]$ (see Example 3.18). Then

$$R\text{Hom}(\mathcal{F}_1 * \mathcal{F}_2, \mathcal{F}_3) = R\text{Hom}(\mathbf{k}_{[0,\infty)}, \mathbf{k}_{[0,\infty)}) = \mathbf{k}$$

and

$$R\text{Hom}(\mathcal{F}_1, \mathcal{H}om^*(\mathcal{F}_2, \mathcal{F}_3)) = R\text{Hom}(\mathbf{k}_{[0,\infty)}, \mathbf{k}_{(-\infty,0)}[1])$$

$$= R\text{Hom}(\mathbf{k}_{[0,\infty)}, \mathbf{k}_{(-\infty,0)})[1] = \mathbf{k}[-1][1] = \mathbf{k}.$$

Corollary 3.3 *For any $\mathcal{F} \in \mathcal{D}(\mathbf{k}_{M \times \mathbb{R}})$, we have the conclusion that $\mathcal{H}om^*$ $(\mathbf{k}_{M \times [0,\infty)}, \mathcal{F}) \in \mathcal{D}_{\{\tau \le 0\}} \mathbf{k}_{M \times [0,\infty)})^{\perp,r}$.*

Proof For any $\mathcal{G} \in \mathcal{D}_{\{\tau \le 0\}}(\mathbf{k}_{M \times [0,\infty)})$,

$$R\text{Hom}(\mathcal{G}, \mathcal{H}om^*(\mathbf{k}_{M \times [0,\infty)}, \mathcal{F})) = R\text{Hom}(\mathcal{G} * \mathbf{k}_{M \times [0,\infty)}, \mathcal{F}) = 0,$$

where the last step comes from Remark 3.9. $\square$

Due to this corollary, we can also view/define $\mathcal{T}(M)$ as $\mathcal{D}_{\{\tau \le 0\}}(\mathbf{k}_{M \times [0,\infty)})^{\perp,r}$, the right orthogonal complement, since $\mathcal{H}om^*(\mathbf{k}_{M \times [0,\infty)}, \cdot)$ also provides an orthogonal decomposition.

Corollary 3.4 *$\mathcal{H}om^*$ is well-defined in $\mathcal{T}(M)$ (view $\mathcal{T}(M)$ as the right orthogonal complement $\mathcal{D}_{\{\tau \le 0\}}(\mathbf{k}_{M \times [0,\infty)})^{\perp,r}$).*

Proof For any $\mathcal{F}, \mathcal{G} \in \mathcal{T}(M)$, for any $\mathcal{H} \in \mathcal{D}_{\{\tau \leq 0\}}(\mathbf{k}_{M \times [0,\infty)})$,

$$RHom(\mathcal{H}, \mathcal{H}om^*(\mathcal{F}, \mathcal{G})) = RHom(\mathcal{H} * \mathcal{F}, \mathcal{G})$$
$$= RHom(\mathcal{H} * \mathbf{k}_{M \times [0,\infty)} * \mathcal{F}, \mathcal{G}) = 0,$$

where the final step comes from the fact that $\mathcal{H} * \mathbf{k}_{M \times [0,\infty)} = 0$, which is explained in Remark 3.9. □

Now, we are ready to prove identity (3.10).

Proof (*of the equality (3.10)*) For any $\mathcal{F}, \mathcal{G} \in \mathcal{T}(M)$, denote by $\pi : M \times \mathbb{R} \to \mathbb{R}$ the projection. Then

$$RHom(\mathcal{F}, \mathcal{G}) = RHom(\mathbf{k}_{M \times [0,\infty)} * \mathcal{F}, \mathcal{G})$$
$$= RHom(\mathbf{k}_{M \times [0,\infty)}, \mathcal{H}om^*(\mathcal{F}, \mathcal{G}))$$
$$= RHom(\pi^{-1}\mathbf{k}_{[0,\infty)}, \mathcal{H}om^*(\mathcal{F}, \mathcal{G})))$$
$$= RHom(\mathbf{k}_{[0,\infty)}, R\pi_*\mathcal{H}om^*(\mathcal{F}, \mathcal{G}))).$$

Besides all the functorial properties used in this argument, the only non-trivial step is the first equality which uses the criterion in Theorem 3.1. □

Now, we are ready to give a proof of the Separation Theorem. This is a good example showing that conditions on the singular support can impose strong restrictions on the behavior the sheaf. Recall the following useful result (Corollary 1.7 in [26] and cf. Proposition 4.2).

Lemma 3.4 *For any* $\mathcal{F} \in \mathcal{D}(\mathbf{k}_{M \times \mathbb{R}})$, *if* $SS(\mathcal{F}) \cap (0_M \times T^*\mathbb{R}) \subset 0_{M \times \mathbb{R}}$, *then* $R\pi_*\mathcal{F}$ *is a constant sheaf. Here* $\pi : M \times \mathbb{R} \to \mathbb{R}$ *is the projection.*

Proof (*Proof of Theorem 3.3*) By identity (3.10) and Lemma 3.4, we aim to show that if both $\mathcal{F}$ and $\mathcal{G}$ are satisfy the hypothesis, then

$$SS(\mathcal{H}om^*(\mathcal{F}, \mathcal{G})) \cap (0_M \times T^*\mathbb{R}) \subset 0_{M \times \mathbb{R}}.$$

By the geometric meaning of the sheaf convolution, we will focus on $SS(\overline{\mathcal{F}}) \hat{*} SS(\mathcal{G})$. Due to Remark 3.16, we can assume that $\mathcal{F}$ and $\mathcal{G}$ have some common M-component with colinear co-vectors. Then

$$SS(\overline{\mathcal{F}}) \hat{*} SS(\mathcal{G}) \subset \{(m, \tau(\xi_{\mathcal{G}} - \xi_{\mathcal{F}}), t_{\mathcal{F}} + t_{\mathcal{G}}, \tau \,|\, \text{common } m \text{ and } \tau\}.$$

Intersecting with $0_M \times T^*\mathbb{R}$, we get $\tau(\xi_{\mathcal{G}} - \xi_{\mathcal{F}}) = 0$. But $A \cap B = \emptyset$ implies that, over this common m, $\xi_{\mathcal{G}} \neq \xi_{\mathcal{F}}$, so we are left that $\tau = 0$. Thus by Lemma 3.4, $R\pi_*\mathcal{H}om^*(\mathcal{F}, \mathcal{G})$ is just a constant sheaf.

Finally, we will check that $\Gamma(\mathbb{R}, R\pi_* \mathcal{H}om^*(\mathcal{F}, \mathcal{G})) = 0$ (hence $R\pi_* \mathcal{H}om^*$ $(\mathcal{F}, \mathcal{G}) = 0$). Indeed,

$$
\begin{aligned}
R\Gamma(\mathbb{R}, R\pi_* \mathcal{H}om^*(\mathcal{F}, \mathcal{G})) &= R\mathrm{Hom}(\mathbf{k}_\mathbb{R}, R\pi_* \mathcal{H}om^*(\mathcal{F}, \mathcal{G})) \\
&= R\mathrm{Hom}(\mathbf{k}_{M \times \mathbb{R}} * \mathcal{F}, \mathcal{G}) \\
&= R\mathrm{Hom}(\mathbf{k}_{M \times \mathbb{R}} * \mathbf{k}_{M \times [0,\infty)} * \mathcal{F}, \mathcal{G}) = 0,
\end{aligned}
$$

where the final step comes from an easy computation that $\mathbf{k}_{M \times \mathbb{R}} * \mathbf{k}_{M \times [0,\infty)} = 0$.

$\square$

3.9 Sheaf Barcodes from Generating Functions

This is a special section aimed at understanding the topological meaning of the sheaf $\mathcal{H} := R\pi_* \mathcal{H}om^*(\mathcal{F}, \mathcal{G})$ that plays a crucial role in the Separation Theorem. The highlight of this section is Theorem 3.4. We will start with the concrete Example 3.19, that is, take $\mathcal{F} = \mathcal{F}_f$ and $\mathcal{G} = \mathcal{F}_g$ for some functions f, g on M. Recall that, for such $\mathcal{F}$ and $\mathcal{G}$,

$$
\mathcal{H}om^*(\mathcal{F}, \mathcal{G}) = \mathcal{H}om^*(\mathcal{F}_f, \mathcal{F}_g) = \mathbf{k}_{\{(m,t) \in M \times \mathbb{R} \mid f(m) - g(m) > t\}}[1].
$$

In this section, we assume that M is compact and $f - g$ is Morse on M.

Topological stalk of $\mathcal{H}$. We can investigate this sheaf $\mathcal{H}$ by first looking at its stalks. For any $t \in \mathbb{R} \backslash \{$critical values of $f - g\}$,

$$
\mathcal{H}_t = (R\pi_* \mathbf{k}_{\{f-g>t\}})_t = H^*(M; \mathbf{k}_{\{f-g>t\}}), \tag{3.13}
$$

up to a degree shift. Since $\{f - g > t\}$ is open, we consider the exact sequence

$$
0 \longrightarrow \mathbf{k}_{\{f-g>t\}} \longrightarrow \mathbf{k}_M \longrightarrow \mathbf{k}_{\{f-g \le t\}} \longrightarrow 0
$$

which after applying $H^*(M, \cdot)$, yields the long exact sequence

$$
H^*(M; \mathbf{k}_{\{f-g>t\}}) \longrightarrow H^*(M; \mathbf{k}_M) \longrightarrow H^*(M; \mathbf{k}_{\{f-g \le t\}}) \xrightarrow{+1}
$$

which coincides with

$$
H^*(M; \mathbf{k}_{\{f-g>t\}}) \longrightarrow H^*(M; \mathbf{k}) \longrightarrow H^*(\{f - g \le t\}; \mathbf{k}) \xrightarrow{+1}
$$

due to the well-known fact that $H^*(M; \mathbf{k}_N) = H^*(N; \mathbf{k})$ if N is a closed submanifold of M. Then by the Five Lemma, $H^*(M; \mathbf{k}_{\{f-g>t\}}) \simeq H^*(M, \{f-g \leq t\}; \mathbf{k})$. On the other hand,

$$H^*(M, \{f - g \leq t\}; \mathbf{k}) = H^*(\{f - g \geq t\}, \partial\{f - g \geq t\}; \mathbf{k}) \quad \text{Excision}$$

$$= H_{n-*}(\{f - g \geq t\}; \mathbf{k}) \quad \text{Lefschetz duality}$$

The version of Lefschetz duality used here is the following: for any pair $(M, \partial M)$, $H^*(M, \partial M; \mathbf{k}) = H_{n-*}(M; \mathbf{k})$. The resulting homology is similar to the generating function homology (GH-homology) defined in [46]. To some extent, we can regard $\mathcal{H}$ as a generalization of GH-homology in the language of sheaves.

Remark 3.18 The computation above only describes the stalks of $\mathcal{H}$ at all but the critical values. In other words, if we view $H_{n-*}(\{f - g \geq t\}; \mathbf{k})$ as a package to form a persistence $\mathbf{k}$-module, we don't know the open-closedness of the endpoints for each corresponding bar in its barcode. However, we are free to choose how to "complete" the endpoints (cf. Theorem 3.4).

Constructibility of $\mathcal{H}$. By (3.11), $\mathcal{H}om^*(\mathcal{F}, \mathcal{G})$ is constructible, so by Proposition 8.4.8 in [32], $\mathcal{H}$ is also constructible. Then (up to degree shifts) $\mathcal{H} = \bigoplus \mathbf{k}_{[a,b)}$. Notice that here we actually know the open-closedness of the endpoints, because we are working in the Tamarkin category. The collection of intervals in this decomposition is called the *sheaf barcode of $\mathcal{H}$*, and is denoted by $\mathcal{B}(\mathcal{H})$.

Without referring to Proposition 8.4.8 in [32], by the standard Morse theory, we still know that $\mathcal{H}$ is constructible since $\mathcal{H}_s \simeq \mathcal{H}_t$ if there are no critical values between s and t, which implies that the endpoints a, b in the bars of $\mathcal{B}(\mathcal{H})$ necessarily lie in the set of critical values of $f - g$. Equivalently, the stalk change only at the critical values of $f - g$, which we denote by $\lambda_1 < \lambda_2 < \cdots < \lambda_n$. Finally, by the microlocal Morse lemma, for any $s < t$, there exists a well-defined map $\tau_{t,s} : \mathcal{H}_t \to \mathcal{H}_s$ (**Exercise**). Thus one gets Figure 3.8, where before λ_1 (the minimal critical value), we have a constant sheaf with rank equal to the rank of $H_*(M; \mathbf{k})$ and after λ_n (the maximal critical value), just zero. In fact, these data *uniquely* determine the decomposition of $\mathcal{H}$ by the following lemma.

Lemma 3.5 *A constructible sheaf $\mathcal{F}$ over $\mathbb{R}$ with $SS(\mathcal{F}) \subset \{\tau \geq 0\}$ is uniquely determined by the following data,[2]*

- *a list of numbers $\{\lambda_0, \lambda_1, \ldots, \lambda_n\} \subset \mathbb{R}$, where $\lambda_0 = -\infty$;*
- *non-negative integers $N_0, \ldots N_n$ such that $\mathcal{F}|_{(\lambda_i, \lambda_{i+1})} = \mathbf{k}^{N_i}$;*
- *morphisms $\tau_{a_{i+1}, a_i} : \mathcal{F}_{a_{i+1}} \to \mathcal{F}_{a_i}$ for $a_i \in (\lambda_i, \lambda_{i+1})$ and $i = 0, \ldots, n$.*

---

[2]This lemma arises from a claim in [24], Section 7. The original claim differs from the one here through the third condition, which says that, to determine a general constructible sheaf $\mathcal{F}$ over $\mathbb{R}$, we need to know morphisms $\mathcal{F}(\lambda_i, \lambda_{i+1}) \leftarrow \mathcal{F}(\lambda_i, \lambda_{i+2}) \to \mathcal{F}(\lambda_{i+1}, \lambda_{i+2})$ for each i. However, since our sheaf $\mathcal{F}$ satisfies $SS(\mathcal{F}) \subset \{\tau \geq 0\}$, by the microlocal Morse lemma, $\mathcal{F}(\lambda_i, \lambda_{i+2}) \simeq \mathcal{F}(\lambda_{i+1}, \lambda_{i+2})$, where λ_{i+1} propagates to λ_i. Therefore, we are reduced to the case $\mathcal{F}(\lambda_{i+1}, \lambda_{i+2}) \to \mathcal{F}(\lambda_i, \lambda_{i+1})$, which is of course equivalent to $\mathcal{F}_{a_{i+1}} \to \mathcal{F}_{a_i}$.

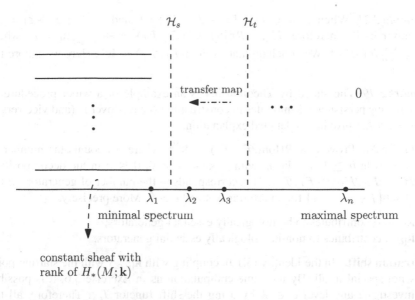

Fig. 3.8 Sheaf barcode derived from generating functions

Be cautious that, without having the information provided by $\tau_{a_i,a_{i+1}}$, $\mathcal{F}$ will not be uniquely determined even if we know the rank N_i on each connected component as well as all the critical values λ_i. Meanwhile, up to all the isomorphisms exhibited earlier, $\tau_{t,s}$ is induced by the map $H_{n-*}(\{f - g > t\}; \mathbf{k}) \to H_{n-*}(\{f - g > s\}; \mathbf{k})$, induced by the inclusion $\{f - g > t\} \hookrightarrow \{f - g > s\}$. Therefore, $\tau_{a_{i+1},a_i} \simeq \iota_{a_{i+1},a_i}$, where ι_{a_{i+1},a_i} is the transfer map for the persistence $\mathbf{k}$-module given by $H_{n-*}(\{f - g \geq t\}; \mathbf{k})$. Since they give the same quiver representation (see Sect. 2.6), the following result holds true.

Theorem 3.4 *There are two ways to generate barcodes of a Morse system* (M, h), *where* $h : M \to \mathbb{R}$ *is a Morse function on the smooth manifold* M.

(1) *(Morse-persistence) Compute the (anti)-persistence* $\mathbf{k}$-*module*

$$V := \left\{ \{H_*(\{h \geq t\}; \mathbf{k})\}_{t \in \mathbb{R}}; \iota_{t,s} \right\}_{s \leq t}$$

for regular values t *and complete its barcode in* $[-, -)$-*type.*

(2) *(Sheaf-constructible) Compute the sheaf (over* $\mathbb{R}$),

$$\mathcal{H} := R\pi_* \mathcal{H}om^*(\mathcal{F}_f, \mathcal{F}_g), \text{ with } f - g = h,$$

where $\pi : M \times \mathbb{R} \to \mathbb{R}$ *is the projection onto the* $\mathbb{R}$-*component. Moreover, after applying decomposition theorems for each case,* $\mathcal{B}(V) = \mathcal{B}(\mathcal{H})$.

Example 3.21 When $f = g$, then $\{f - g > 0\} = \emptyset$ and $\{f - g > -\epsilon\} = M$ for any $\epsilon > 0$. Therefore, $\mathcal{H} := R\pi_* \mathcal{H}om^*(\mathcal{F}_f, \mathcal{F}_f) = \bigoplus_{N \text{ copies}} \mathbf{k}_{(-\infty,0)}$, where $N = \sum_i b_i(M; \mathbf{k})$. We conclude that, $R\mathrm{Hom}(\mathcal{F}_f, \mathcal{F}_f) = \mathbf{k}^N$. Here we ignore the degrees.

Remark 3.19 The stated by Theorem 3.4 is an example of a wider procedure of transferring persistence $\mathbf{k}$-modules to constructible sheaves over $\mathbb{R}$ (and vice versa). Appendix A.1 provides a detailed explanation.

Exercise 3.6 Prove that $R\mathrm{Hom}(\mathcal{F}_f, \mathcal{F}_g) = \mathbf{k}^{N_0}$, where N_0 counts the number of $\mathbf{k}_{(-\infty,b)}$ with $b \geq 0$ and $\mathbf{k}_{[a,b)}$ with $-\infty < a < 0, 0 \leq b$ in the decomposition of $\mathcal{H} := R\pi_* \mathcal{H}om^*(\mathcal{F}_f, \mathcal{F}_g)$. This corresponds to the number of generators *at the* level set $\{f - g = -\epsilon\}$ for an arbitrarily small $\epsilon > 0$. More precisely,

- $\mathbf{k}_{(-\infty,b)}$ contributes to homologically essential generators;
- $\mathbf{k}_{[a,b)}$ contributes to non-homologically essential generators.

Filtration shift. In the identity (3.10), coupling with $\mathbf{k}_{[0,\infty)}$ with the starting point 0 is not special at all. By the same computation as in Exercise 3.6, it is possible to investigate any level $c \in \mathbb{R}$ by using the shift functor T_{c*}. Therefore, all the Morse information (including the intermediate born-killed relations on generators) can be recovered. Explicitly, for any $\mathcal{F}, \mathcal{G} \in \mathcal{T}(M)$ and for any $c \in \mathbb{R}$, we have the following filtration-shifted version of the computation of $R\mathrm{Hom}(\mathcal{F}, \mathcal{G})$:

$$
\begin{aligned}
R\mathrm{Hom}(\mathcal{F}, T_{c*}\mathcal{G}) &= R\mathrm{Hom}(T_{-c*}\mathcal{F}, \mathcal{G}) \\
&= R\mathrm{Hom}(\mathbf{k}_{M\times[-c,\infty)} * \mathcal{F}, \mathcal{G}) \\
&= R\mathrm{Hom}(\mathbf{k}_{M\times[-c,\infty)}, \mathcal{H}om^*(\mathcal{F}, \mathcal{G})) \\
&= R\mathrm{Hom}(\pi^{-1}\mathbf{k}_{[-c,\infty)}, R\pi_*\mathcal{H}om^*(\mathcal{F}, \mathcal{G})) \\
&= R\mathrm{Hom}(\mathbf{k}_{[-c,\infty)}, R\pi_*\mathcal{H}om^*(\mathcal{F}, \mathcal{G})) \\
&= R\mathrm{Hom}(\mathbf{k}_{[0,\infty)}, T_{c*}R\pi_*\mathcal{H}om^*(\mathcal{F}, \mathcal{G})).
\end{aligned}
$$

In the case of Morse functions, we will get $\mathbf{k}^{N_c}$, where N_c counts the number of generators appearing at the level set $\{f - g = -c - \epsilon\}$ for an arbitrarily small $\epsilon > 0$. N_c changes only at the critical values c.

Remark 3.20 Compared with persistent homology, the pair $\{(c, N_c)\}$ exactly corresponds to the persistence Betti number β_c. Moreover, for any $c < d$, the well-defined map $\tau_{c,d} : T_{c*}\mathcal{G} \to T_{d*}\mathcal{G}$ induces a well-defined map $\iota_{c,d} : R\mathrm{Hom}(\mathcal{F}, T_{c*}\mathcal{G}) \to R\mathrm{Hom}(\mathcal{F}, T_{d*}\mathcal{G})$. Thus the following direct system

$$ W := \{\{R\mathrm{Hom}(\mathcal{F}, T_{c*}\mathcal{G})\}_{c\in\mathbb{R}}, \iota_{c,d}\} $$

forms a persistence $\mathbf{k}$-module. Moreover, $\mathcal{B}(W) = \mathcal{B}(V) = \mathcal{B}(\mathcal{H})$ in Theorem 3.4 (modulo the endpoints because $\mathcal{B}(W)$ gives opposite open-closed bars compared with $\mathcal{B}(V)$ or $\mathcal{B}(\mathcal{H})$). Later, in Sect. 4.3, we will see how the "torsions" of $\mathcal{H}$, which are equivalent to the lengths of finite length bars in $\mathcal{B}(\mathcal{H})$, can capture the displacement energies.

3.10 Interleaving Distance in a Tamarkin Category

In this section, we will define an interleaving relation between two objects in a Tamarkin category, and provide a criterion that allows one to check how much two objects are interleaved.

3.10.1 Definition of Sheaf Interleaving

Recall that, in $\mathcal{T}(M)$ and for any $c \geq 0$, there exists a natural transformation τ_c : $\mathbb{1} \to T_{c*}$.

Definition 3.11 Two sheaves $\mathcal{F}, \mathcal{G} \in \mathcal{T}(M)$ are said to be *c-interleaved* ($c \geq 0$) if there exist (in general, two pairs of maps) $\mathcal{F} \xrightarrow{\alpha,\delta} T_{c*}\mathcal{G}$ and $\mathcal{G} \xrightarrow{\beta,\gamma} T_{c*}\mathcal{F}$ such that the following two diagrams commute:

$$\mathcal{F} \xrightarrow{\alpha} T_{c*}\mathcal{G} \xrightarrow{T_{c*}\beta} T_{2c*}\mathcal{F}$$
$$\tau_{2c}(\mathcal{F})$$

and

$$\mathcal{G} \xrightarrow{\gamma} T_{c*}\mathcal{F} \xrightarrow{T_{c*}\delta} T_{2c*}\mathcal{G} .$$
$$\tau_{2c}(\mathcal{G})$$

In this case we can define a distance by

$$d_{\mathcal{T}(M)}(\mathcal{F}, \mathcal{G}) = \inf\{c \geq 0 \mid \mathcal{F} \text{ and } \mathcal{G} \text{ are } c\text{-interleaved}\}. \tag{3.14}$$

Remark 3.21 (1) $\mathcal{F}$ is 2c-torsion if and only if $\mathcal{F}$ and 0 are c-interleaved. (2) When $M = \{pt\}$, this gives a distance similar to d_{int} for persistence **k**-modules (see Sect. 2.7). However, due to the non-symmetric pairs, (3.14) defines a smaller distance than the actual interleaving distance. (3) In [1], an (a, b)-isometry relation was introduced that is a non-balanced version of our definition given above. Obviously, (a, b)-isometry implies $(a + b)$-interleaved.

Example 3.22 $\mathbf{k}_{[0,\infty)}$ and $\mathbf{k}_{[2,\infty)}$ are 2-interleaved.

Fig. 3.9 Differentiable
function f on the circle S^1

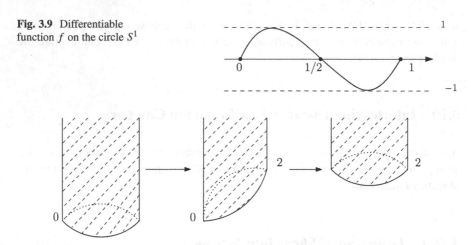

Fig. 3.10 An example of sheaf interleaving

Example 3.23 Let $M = S^1 = \mathbb{R}/\mathbb{Z}$ and let $f : S^1 \to \mathbb{R}$ be the function show in Figure 3.9. Consider the canonical sheaf $\mathcal{F}_f$ of f, defined in Example 3.9. It is easy to see that $\mathcal{F}_f$ and $\mathbf{k}_{M \times [0,\infty)}$ are 1-interleaved. This is depicted in Figure 3.10. In fact, in this case, $d_{\mathcal{T}(S^1)}(\mathcal{F}_f, \mathbf{k}_{M \times [0,\infty)}) = 1$.

Example 3.24 One can generalize the example above. For any two functions $f, g : M \to \mathbb{R}$ (assuming M is compact) with mean value equal to 0, $d_{\mathcal{T}(M)}(\mathcal{F}_f, \mathcal{F}_g) \le |\max_M f - \min_M g|$.

Proposition 3.7 (Exercise) *We list some basic properties of $d_{\mathcal{T}(M)}$.*

(1) $d_{\mathcal{T}(M)}(\cdot, \cdot)$ *is a pseudo-metric on $\mathcal{T}(M)$.*
(2) $d_{\mathcal{T}(M)}(\mathcal{F}_1, \mathcal{F}_2) = d_{\mathcal{T}(M)}(\bar{\mathcal{F}}_1, \bar{\mathcal{F}}_2)$.
(3) $d_{\mathcal{T}(M)}(\mathcal{F}_1 * \mathcal{G}_1, \mathcal{F}_2 * \mathcal{G}_2) \le d_{\mathcal{T}(M)}(\mathcal{F}_1, \mathcal{F}_2) + d_{\mathcal{T}(M)}(\mathcal{G}_1, \mathcal{G}_2)$.
(4) $d_{\mathcal{T}(M)}(\mathcal{F}_1 *_{np} \mathcal{G}_1, \mathcal{F}_2 *_{np} \mathcal{G}_2) \le d_{\mathcal{T}(M)}(\mathcal{F}_1, \mathcal{F}_2) + d_{\mathcal{T}(M)}(\mathcal{G}_1, \mathcal{G}_2)$.
(5) $d_{\mathcal{T}(M)}(\mathcal{H}om^*(\mathcal{F}_1, \mathcal{G}_1), \mathcal{H}om^*(\mathcal{F}_2, \mathcal{G}_2)) \le d_{\mathcal{T}(M)}(\mathcal{F}_1, \mathcal{F}_2) + d_{\mathcal{T}(M)}(\mathcal{G}_1, \mathcal{G}_2)$.
(6) $d_{\mathcal{T}(M)}(R\pi_* \mathcal{F}_1, R\pi_* \mathcal{F}_2) \le d_{\mathcal{T}(M)}(\mathcal{F}_1, \mathcal{F}_2)$, *with $\pi : M \times \mathbb{R} \to \mathbb{R}$ the projection.*

Remark 3.22 Example 3.24 is simple, yet enlightening in the following sense. It shows that the pseudo-metric $d_{\mathcal{T}(M)}$ on sheaves is related to the C^0-distance of generating functions that define Lagrangian submanifolds. Not every Lagrangian submanifold admits a generating function, but sometimes it can be C^0-approximated by a family of Lagrangian submanifolds that do admit generating functions. Figure 3.11 provides a standard example on $T^*\mathbb{R}$. Suggested by L. Polterovich, the sheaf method can be used to study this C^0-phenomenon in symplectic geometry. Our first intuition comes from the following heuristic deduction that, depending on the metric properties of space $(\mathcal{T}(M), d_{\mathcal{T}(M)})$, which we conjecture it is a complete metric space, such a C^0-approximation of Lagrangian submanifolds provide a Cauchy sequence in $\mathcal{T}(M)$ with respect to the pseudo-metric $d_{\mathcal{T}(M)}$. Then

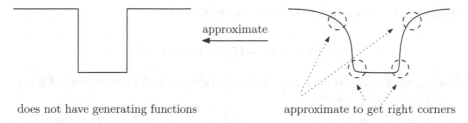

does not have generating functions   approximate to get right corners

Fig. 3.11 C^0-approximation of a bad Lagrangian

one obtains a *limit sheaf* in $\mathcal{T}(M)$ representing (via SS) this "bad" Lagrangian submanifold like in Figure 3.11. Importantly, this limit sheaf is *not* from any of those that can be constructed from generating functions! Then instead of studying this Lagrangian submanifold directly (which might be hard or impossible by some classical methods), we can switch to exploring algebraic properties of its associated limit sheaf and thus apply microlocal sheaf theory. Note that in [23] the microlocal sheaf method was successfully used in C^0-symplectic geometry to reprove the Gromov-Eliashberg Theorem.

3.10.2 Torsion Criterion

We start this subsection with the following question: Given two sheafs $\mathcal{F}, \mathcal{G} \in \mathcal{T}(M)$, how do we check whether or not they are $(c\text{-})$interleaved? Here is a useful criterion, which we call *torsion criterion*. Recall that, by Remark 3.14, a torsion element can be defined in the category $\mathcal{D}_{\{\tau \geq 0\}}(M \times \mathbb{R})$, so let us state this criterion in this set-up.

Theorem 3.5 *Let $g : \mathcal{G} \to \mathcal{H}$ be a morphism. If $\mathcal{G} \xrightarrow{g} \mathcal{H}$ completes to be a distinguished triangle $\mathcal{F} \xrightarrow{f} \mathcal{G} \xrightarrow{g} \mathcal{H} \xrightarrow{+1}$ in $\mathcal{D}_{\{\tau \geq 0\}}(M \times \mathbb{R})$ such that $\mathcal{F}$ is c-torsion, then $\mathcal{G}$ and $\mathcal{H}$ are c-interleaved.*

Motivation. Let us consider $\mathcal{D}(\mathcal{P})$, the (bounded) derived category of persistence **k**-modules. We can identify each $V \in \mathcal{P}$ with a (bounded) complex $V_\bullet := (\cdots 0 \to V \to 0 \cdots)$. Obviously, any morphism $f : V \to W$ is then identified with a chain map $f_\bullet = (\ldots, 0, f, 0, \ldots)$. The category $\mathcal{D}(\mathcal{P})$ is naturally a triangulated category and hence $f_\bullet : V_\bullet \to W_\bullet$ can be always completed by adding $\text{Cone}(f)_\bullet$, where

$$\text{Cone}(f)_1 = W, \quad \text{Cone}(f)_0 = V \quad \text{and} \quad \partial_{\text{co}} = f : \text{Cone}(f)_0 \to \text{Cone}(f)_1.$$

Since the homological dimension of $\mathcal{G}$ is equal to 1, by Exercise I.18 in [32], every complex $X_\bullet \in \mathcal{D}(\mathcal{P})$ has the property that $X_\bullet \simeq \bigoplus_i H^i(X)[-i]$. In particular,

$$\text{Cone}(f)_\bullet \simeq \text{coker}(f)[-1] \oplus \ker(f)$$

Therefore, we get a distinguished triangle in $\mathcal{D}(\mathcal{P})$,

$$V \xrightarrow{f} W \longrightarrow \operatorname{coker}(f)[-1] \oplus \ker(f) \xrightarrow{+1} .$$

Finally, we remark that the same construction also works for the category of finite-dimensional vector spaces.

Example 3.25 Let $V = \mathbb{I}_{(5,\infty)} \oplus \mathbb{I}_{(2,8]}$ and $W = \mathbb{I}_{(3,\infty)} \oplus \mathbb{I}_{(0,6]}$. Then the natural "identity map" f, i.e., identity on non-trivial part, gives

$$\ker(f) = \mathbb{I}_{(6,8]} \quad \text{and} \quad \operatorname{coker}(f) = \mathbb{I}_{(3,5]} \oplus \mathbb{I}_{(0,2]}.$$

Note that $\operatorname{coker}(f)[-1] \oplus \ker(f)$ is 2-torsion, which corresponds to the fact that V and W are 2-interleaved. In general, if $\operatorname{coker}(f)[-1] \oplus \ker(f)$ are torsion, then the barcodes of V and W will have the same number of infinite-length bars, and also have almost the same finite-length bars up to some shifts at endpoints depending on the torsions.

The proof of Theorem 3.5 essentially relies on the following well-known lemma.

Lemma 3.6 *In a derived category,* $\operatorname{Hom}(A, \cdot)$ *and* $\operatorname{Hom}(\cdot, B)$ *are cohomological functors, i.e., for any distinguished triangle* $X \longrightarrow Y \longrightarrow Z \xrightarrow{+1}$,

$$\operatorname{Hom}(A, X) \longrightarrow \operatorname{Hom}(A, Y) \longrightarrow \operatorname{Hom}(A, Z) \xrightarrow{+1}$$

and

$$\operatorname{Hom}(X, B) \longrightarrow \operatorname{Hom}(Y, B) \longrightarrow \operatorname{Hom}(Z, B) \xrightarrow{+1}$$

are long exact sequences.

Instead of giving the proof of Lemma 3.6 (which can be derived from item (ii) in Proposition 1.5.3 in [32]), we leave it as a good exercise to enhance the familiarity with various axioms of a triangulated category. Moreover, we provide an example to demonstrate this lemma.

Example 3.26 Let V, W be two vector spaces. Let $f : V \to W$ be an injective map. Complete this to be a distinguished triangle in the derived category of vector spaces,

$$\cdots \longrightarrow V \xrightarrow{f} W \xrightarrow{i} \operatorname{coker}(f) \xrightarrow{0} 0 \longrightarrow \cdots$$

Applying $\operatorname{Hom}(V, \cdot)$ to this distinguished triangle, we get

$$\cdots \longrightarrow \operatorname{Hom}(V, V) \xrightarrow{f\circ} \operatorname{Hom}(V, W) \xrightarrow{i\circ} \operatorname{Hom}(V, \operatorname{coker}(f)) \xrightarrow{0} 0 \longrightarrow \cdots$$

An elementary argument shows that the long sequence above is exact. For instance, for any $\phi \in \mathrm{Hom}(V, W)$, the equality $i \circ \phi = 0$ is equivalent to $\phi(V) \subset \mathrm{im}(f)$. For any $v \in V$, by injectivity of f, there exists a unique $v' \in V$ such that $\phi(v) = f(v')$. Define $\psi \in \mathrm{Hom}(V, V)$ by $\psi(v) = v'$. Then $f \circ \psi = \phi$.

Proof *(of Theorem 3.5)* By the definition of a torsion element, we have the diagram

$$
\begin{array}{ccccccc}
\mathcal{F} & \xrightarrow{f} & \mathcal{G} & \xrightarrow{g} & \mathcal{H} & \xrightarrow{h} & \mathcal{F}[1] \\
\downarrow{\scriptstyle 0} & & \downarrow{\scriptstyle \tau_c(\mathcal{G})} & & \downarrow{\scriptstyle \tau_c(\mathcal{H})} & & \downarrow{\scriptstyle 0} \\
T_{c*}\mathcal{F} & \xrightarrow{T_{c*}f} & T_{c*}\mathcal{G} & \xrightarrow{T_{c*}g} & T_{c*}\mathcal{H} & \xrightarrow{T_{c*}h} & T_{c*}\mathcal{F}[1]
\end{array}
$$

Applying $\mathrm{Hom}(\mathcal{H}, \cdot)$ to the bottom row, we get the long exact sequence

$$
\cdots \to \mathrm{Hom}(\mathcal{H}, T_{c*}\mathcal{G}) \xrightarrow{T_{c*}g \circ} \mathrm{Hom}(\mathcal{H}, T_{c*}\mathcal{H}) \xrightarrow{T_{c*}h \circ} \mathrm{Hom}(\mathcal{H}, T_{c*}\mathcal{F}[1]) \to \cdots
$$

Since $T_{c*}h \circ \tau_c(\mathcal{H}) = 0$, there exists some $\beta : \mathcal{H} \to T_{c*}\mathcal{G}$ such that $T_{c*}g \circ \beta = \tau_c(\mathcal{H})$. Hence,

$$
\mathcal{H} \xrightarrow{\beta} T_{c*}\mathcal{G} \xrightarrow{T_{c*}g} T_{c*}\mathcal{H} \xrightarrow{\tau_{c,2c}(\mathcal{H})} T_{2c*}\mathcal{H}
$$
$$
\underset{\tau_{2c}(\mathcal{H})}{\underbrace{}}
$$

Similarly, applying $\mathrm{Hom}(\cdot, T_{c*}\mathcal{G})$ to the top line, we will get a long exact sequence

$$
\cdots \longrightarrow \mathrm{Hom}(\mathcal{H}, T_{c*}\mathcal{G}) \xrightarrow{\circ g} \mathrm{Hom}(\mathcal{G}, T_{c*}\mathcal{G}) \xrightarrow{\circ f} \mathrm{Hom}(\mathcal{F}, T_{c*}\mathcal{G}) \longrightarrow \cdots
$$

Since $\tau_c(\mathcal{G}) \circ f = 0$, there exists some $\gamma : \mathcal{H} \to T_{c*}\mathcal{G}$ such that $\gamma \circ g = \tau_c(\mathcal{G})$. Hence, $T_{c*}(\gamma \circ g) = T_{c*}\gamma \circ T_{c*}g = \tau_{c,2c}(\mathcal{G})$, and so

$$
\mathcal{G} \xrightarrow{\tau_c(\mathcal{G})} T_{c*}\mathcal{G} \xrightarrow{T_{c*}g} T_{c*}\mathcal{H} \xrightarrow{T_{c*}\gamma} T_{2c*}\mathcal{G}
$$
$$
\underset{\tau_{2c}(\mathcal{G})}{\underbrace{}}
$$

Compared with Definition 3.11, $\alpha = \tau_c(\mathcal{H}) \circ g$ and $\delta = T_{c*}g \circ \tau_c(\mathcal{G})$ are morphisms from $\mathcal{G}$ to $T_{c*}\mathcal{H}$; β, γ that come from the exactness of long exact sequences above are morphisms from $\mathcal{H}$ to $T_{c*}\mathcal{G}$. $\qquad \square$

Remark 3.23 Note that in the proof of Theorem 3.5, the morphisms forming an interleaving relation satisfy $\alpha = \delta$. This enable us to upgrade Definition 3.11 from two pairs of maps to the "1.5-pair" case, i.e., there exists morphisms $\mathcal{F} \xrightarrow{\alpha} T_{c*}\mathcal{G}$ and $\mathcal{G} \xrightarrow{\beta, \gamma} T_{c*}\mathcal{F}$ such that the interleaving relations hold. However, the distance $d_{\mathcal{T}(M)}$ defined in this way is not obviously symmetric (and in fact it might not be symmetric).

The reason why we mention this is that for the category of persistence **k**-modules (equivalently, $\mathcal{T}(\text{pt})$ plus constructibility condition), a careful examination of the proof of the equality $d_{\text{int}} = d_{\text{bottle}}$, shows that this "1.5-pair" is sufficient to obtain the Isometry Theorem. Interestingly, this also implies that in the setting of persistence **k**-modules the interleaving distance for "1.5-pair" morphisms is symmetric because d_{bottle} is symmetric.

3.11 Examples of Interleaving Based on the Torsion Criterion

In this section, we will demonstrate how to use criterion Theorem 3.5 via an elementary but enlightening example.

Denote the coordinates in $\mathbb{R}^2$ by (t, s) and the co-vectors by (τ, σ). Let $\mathcal{F} = \mathbf{k}_T$, where $T = \{(t, s) \in \mathbb{R}^2 \mid -t \leq s \leq t, t \geq 0\}$, see Figure 3.12. We can compute $SS(\mathcal{F})$. There are four cases. Denote by $SS(\mathcal{F})_{(t,s)}$ the fiber of $SS(\mathcal{F})$ over the point (t, s). Then

- when $(t, s) \in \text{int}(T)$, $SS(\mathcal{F})_{(t,s)} = 0$;
- when $t = s(\neq 0)$, $SS(\mathcal{F})_{(t,s)} = \{(\tau, \sigma) \mid \tau = -\sigma, \tau \geq 0\}$;
- when $t = -s(\neq 0)$, $SS(\mathcal{F})_{(t,s)} = \{(\tau, \sigma) \mid \tau = \sigma, \tau \geq 0\}$;
- when $t = s = 0$, $SS(\mathcal{F})_{(t,s)} = \{(\tau, \sigma) \mid -\tau \leq \sigma \leq \tau, \tau \geq 0\}$.

The second and third item come from Proposition 3.3, where $\phi : \mathbb{R}^2 \to \mathbb{R}$ is defined by $\phi(t, s) = t - s$ and $\phi(t, s) = t + s$ respectively. The fourth item comes from Proposition 5.3.1 in [32] saying that $SS(\mathbf{k}_T)_{(0,0)} = T^\circ$ where, viewing T as a closed cone in $\mathbb{R}^2$, T° is its polar cone, defined by $\{(\tau, \sigma) \mid t\tau + s\sigma \geq 0, \ \forall(t, s) \in T\}$. This is depicted in Figure 3.13. Let us point out that

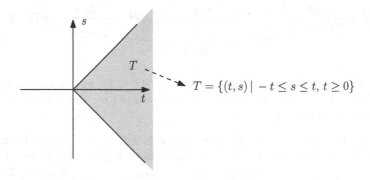

Fig. 3.12 An example of a constant sheaf over a cone

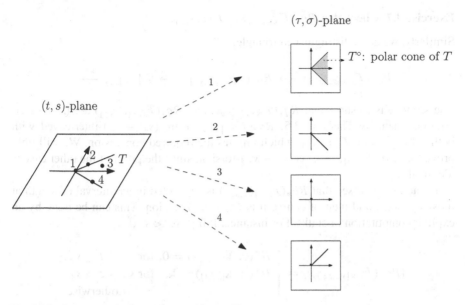

Fig. 3.13 Computation of the singular support of $\mathbf{k}_{\text{cone}}$

$$SS(\mathcal{F}) \subset \{(t, s, \tau, \sigma) \mid -\tau \leq \sigma \leq \tau, \tau \geq 0\} \subset \mathcal{D}_{\{\tau \geq 0\}}(\mathbb{R} \times \mathbb{R}). \qquad (3.15)$$

Then we have the following lemma.

Lemma 3.7 *For any $s_- < s_+$ in $\mathbb{R}$ (in the s-coordinate), $\mathcal{F}|_{\mathbb{R} \times \{s_-\}}$ and $\mathcal{F}|_{\mathbb{R} \times \{s_+\}}$ are $2(s_+ - s_-)$-interleaved.*

Remark 3.24 In fact, since we know $\mathcal{F} = \mathbf{k}_T$, $\mathcal{F}|_{\mathbb{R} \times \{s_-\}} = \mathbf{k}_{[|s_-|,\infty)}$ and $\mathcal{F}|_{\mathbb{R} \times \{s_+\}} = \mathbf{k}_{[|s_+|,\infty)}$. Of course they are at most $(s_+ - s_-)$-interleaved (which is better than what the lemma above states). However, we provide a proof of Lemma 3.7 based on Theorem 3.5, which is more inspiring when dealing with general cases.

Proof (*of Lemma 3.7*) First, we have a short exact sequence

$$0 \longrightarrow \mathcal{F}_{\mathbb{R} \times [s_-, s_+)} \longrightarrow \mathcal{F}_{\mathbb{R} \times [s_-, s_+]} \longrightarrow \mathcal{F}_{\mathbb{R} \times \{s_+\}} \longrightarrow 0$$

Applying the functor Rq_*, where $q : \mathbb{R}^2 \to \mathbb{R}$ by $(t, s) \to t$, we get a distinguished triangle

$$Rq_*(\mathcal{F}_{\mathbb{R} \times [s_-, s_+)}) \longrightarrow Rq_*(\mathcal{F}_{\mathbb{R} \times [s_-, s_+]}) \longrightarrow Rq_*(\mathcal{F}_{\mathbb{R} \times \{s_+\}}) \xrightarrow{+1} .$$

Exercise 3.7 Check that $Rq_*(\mathcal{F}_{\mathbb{R}\times\{s_+\}}) \simeq \mathcal{F}|_{\mathbb{R}\times\{s_+\}}$.

Similarly, we get a distinguished triangle,

$$Rq_*(\mathcal{F}_{\mathbb{R}\times(s_-,s_+]}) \longrightarrow Rq_*(\mathcal{F}_{\mathbb{R}\times[s_-,s_+]}) \longrightarrow \mathcal{F}|_{\mathbb{R}\times\{s_-\}} \xrightarrow{+1} .$$

The strategy is to show that $Rq_*(\mathcal{F}_{\mathbb{R}\times[s_-,s_+]})$ and $Rq_*(\mathcal{F}_{\mathbb{R}\times(s_-,s_+]})$ are $(s_+ - s_-)$-torsion. Then, by Theorem 3.5, $Rq_*(\mathcal{F}_{\mathbb{R}\times[s_-,s_+]})$ are $(s_+ - s_-)$-interleaved with both $\mathcal{F}|_{\mathbb{R}\times\{s_+\}}$ and $\mathcal{F}|_{\mathbb{R}\times\{s_-\}}$, which implies the required conclusion. We will only prove that $Rq_*(\mathcal{F}_{\mathbb{R}\times[s_-,s_+]})$ is $(s_+ - s_-)$-torsion, since the proof of the other case is identical.

In fact, we will see that $Rq_*(\mathcal{F}_{\mathbb{R}\times[s_-,s_+]})$ is supported in an interval of length at most $s_+ - s_-$, and then of course it is $(s_+ - s_-)$-torsion. This can be done by an explicit computation of stalks. For instance, If $0 \leq s_- < s_+$,

$$(Rq_*(\mathcal{F}_{\mathbb{R}\times[s_-,s_+]}))_t = \begin{cases} H^*(\mathbb{R}, \mathbf{k}_{[s_-,s_+)}) = 0, & \text{for} \quad t \geq s_+, \\ H^*(\mathbb{R}, \mathbf{k}_{[s_-,t]}) = \mathbf{k}, & \text{for } s_- \leq t < s_+, \\ 0, & \text{otherwise.} \end{cases}$$

This computation is shown in Figure 3.14. For the other two cases, that is $s_- < 0 \leq s_+$ and $s_- < s_+ < 0$, we get similar results as above, see Figure 3.15. Thus, we described the support of $Rq_*(\mathcal{F}_{\mathbb{R}\times[s_-,s_+]})$ in each case, and this confirms the bounded support conclusion. $\square$

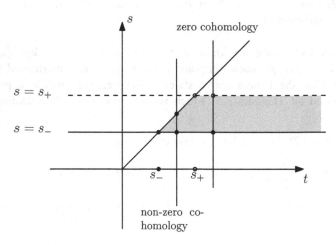

Fig. 3.14 Computation of compactly supported cohomology I

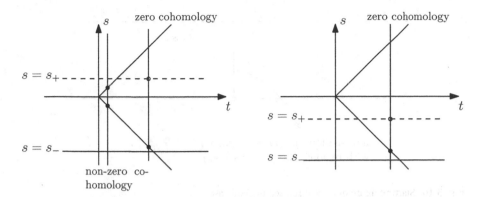

Fig. 3.15 Computation of compactly supported cohomology II

Note that in the proof above, the explicit expression of $\mathcal{F}$ is only used in the final step. In general, we have a stronger result as follows.

Proposition 3.8 *If* $\mathcal{F} \in \mathcal{D}(\mathbf{k}_{\mathbb{R}^2})$ *has the property* (3.15), *then for any* $s_- < s_+$ *in* $\mathbb{R}$ *(in the s coordinate),* $\mathcal{F}|_{\mathbb{R} \times \{s_-\}}$ *and* $\mathcal{F}|_{\mathbb{R} \times \{s_+\}}$ *are* $2(s_+ - s_-)$-*interleaved.*

Based on the proof of Lemma 3.7, we only need to show that $Rq_*(\mathcal{F}_{\mathbb{R} \times [s_-, s_+)})$ is $(s_+ - s_-)$-torsion as long as $\mathcal{F}$ satisfies the condition (3.15) on its singular support. This is another good example showing that conditions on the singular support can impose a strong restriction on the behavior a sheaf (cf. proof of the Separation Theorem in Sect. 3.8). We will give a heuristic proof of this conclusion; the interested reader can check Proposition 5.9 in [27] for a rigorous and detailed proof.

Proof *(not rigorous!)* For $t \in \mathbb{R}$, the stalk at t is

$$(Rq_*(\mathcal{F}_{\mathbb{R} \times [s_-, s_+)}))_t = H^*(\mathbb{R}, \mathcal{F}_{\mathbb{R} \times [s_-, s_+)}|_{\{t\} \times \mathbb{R}}).$$

The only case that this is different from zero is when $\mathcal{F}_{\mathbb{R} \times [s_-, s_+)}|_{\{t\} \times \mathbb{R}}$ has $\mathbf{k}_{[\lambda_-, \lambda_+]}$ or $\mathbf{k}_{(\lambda_-, \lambda_+)}$ as a direct summand. For brevity, we only consider the case where $\mathcal{F}_{\mathbb{R} \times [s_-, s_+)}|_{\{t\} \times \mathbb{R}} = \mathbf{k}_{[\lambda_-, \lambda_+]}$. Without loss of generality, we assume that $\lambda_- = s_-$, and thus get Figure 3.16.

Now, we want to move t and investigate how the sheaf looks like nearby. We list the following four possibilities, see Figure 3.17. The first and last ones are *not* allowed due to the condition (3.15) on the singular support. For the remaining admissible cases, one knows that $\mathcal{F}$ has to "escape" from the strip region (see Figure 3.18). Therefore, identifying $Rq_*(\mathcal{F}_{\mathbb{R} \times [s_-, s_+)})$ with a collection of bars, it does not have any bar with length greater than $s_+ - s_-$, which means it is $(s_+ - s_-)$-torsion. $\square$

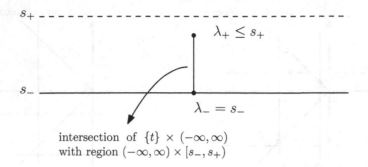

$$s_+$$

$$\lambda_+ \leq s_+$$

$$s_-$$

$$\lambda_- = s_-$$

intersection of $\{t\} \times (-\infty, \infty)$
with region $(-\infty, \infty) \times [s_-, s_+)$

Fig. 3.16 Starting figure of a sheaf (closed interval case)

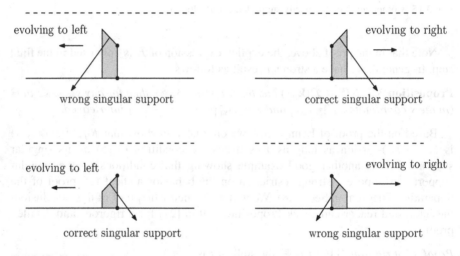

evolving to left

wrong singular support

evolving to right

correct singular support

evolving to left

correct singular support

evolving to right

wrong singular support

Fig. 3.17 Sheaf's evolution constrained by its singular support

Remark 3.25 Note that the proof above provides just a rather sketchy picture, since (i) there might exist various similar pictures with respect to the change of slopes during the evolution process; (ii) since $H^*(\mathbb{R}, \mathbf{k}_{(\lambda_-, \lambda_+)})$ is also non-zero, there might exist evolutions with mixed closed and open interval types. But a careful list of four cases with the open intervals illustrates the same escaping behavior.

Using the same argument, we can prove a general statement.

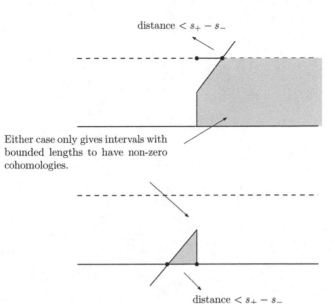

distance $< s_+ - s_-$

Either case only gives intervals with bounded lengths to have non-zero cohomologies.

distance $< s_+ - s_-$

Fig. 3.18 Sheaf evolves and escapes the region

Exercise 3.8 Modify the condition (3.15) to

$$SS(\mathcal{F}) \subset \{(t, s, \tau, \sigma) \mid -b \cdot \tau \leq \sigma \leq a \cdot \tau, \tau \geq 0\} \tag{3.16}$$

for some $a, b > 0$. Then for any $s_- < s_+$ in $\mathbb{R}$ (in the s coordinate), $\mathcal{F}|_{\mathbb{R} \times \{s_-\}}$ and $\mathcal{F}|_{\mathbb{R} \times \{s_+\}}$ are $(a + b)(s_+ - s_-)$-interleaved.

Chapter 4
Applications in Symplectic Geometry

Abstract In this chapter, we discuss various applications of Tamarkin categories in symplectic geometry. We start with a presentation of the Guillermou-Kashiwara-Schapira sheaf quantization, which associates to a homogeneous Hamiltonian diffeomorphism a complex of sheaves with a certain geometric constraint. Next, following the work of Asano and Ike, we establish a stability result with respect to the Hofer norm. Explicitly, the interleaving distance (defined in a Tamarkin category in the previous chapter) provides a lower bound of the Hofer norm. This is followed by an interesting application to displacement energies of subsets of a cotangent bundle. Further, following Chiu's work, we study in detail a restrictive Tamarkin category associated to an open domain U of a Euclidean space. The core of this subject lies in a concept called U-projector. There is a section, from a joint work with Leonid Polterovich, devoted to a geometric interpretation of a U-projector. Finally, after defining a sheaf invariant of a domain, we give a quick proof of Gromov's non-squeezing theorem.

4.1 Guillermou-Kashiwara-Schapira Sheaf Quantization

4.1.1 Statement and Corollaries

The main result in [26], roughly speaking, transfers the study of (homogenous) Hamiltonian isotopies on $\dot{T}^*M$ (here "·" on the top indicates that the zero-section is removed) to a study of sheaves. First, let us consider the following well-known geometric construction.

Definition 4.1 Let $\Phi = \{\phi_t\}_{t \in I}$ be a homogeneous Hamiltonian isotopy on $\dot{T}^*M$ generated by a (homogeneous) Hamiltonian function $F : I \times \dot{T}^*M \to \mathbb{R}$, where $I \subset \mathbb{R}$ is an open interval containing $[0, 1]$. Consider the set

$$\Lambda_\Phi = \left\{ (z, -\phi_t(z), t, -F_t(\phi_t(z))) \,\middle|\, z \in \dot{T}^*M \right\}$$

© Springer Nature Switzerland AG 2020
J. Zhang, *Quantitative Tamarkin Theory*, CRM Short Courses,
https://doi.org/10.1007/978-3-030-37888-2_4

which is a conic Lagrangian submanifold of $\dot{T}^*M \times \dot{T}^*M \times T^*I$, called the *Lagrangian suspension* of Hamiltonian isotopy Φ.

Here is the main result of [26].

Theorem 4.1 *Let M be a compact manifold and $\Phi = \{\phi_t\}_{t \in I}$ be a homogeneous Hamiltonian isotopy on $\dot{T}^*M$, compactly supported[1], where $I \subset \mathbb{R}$ is an open interval containing $[0, 1]$. Then there exists an element $\mathcal{K} \in \mathcal{D}(\mathbf{k}_{M \times M \times I})$ such that*

(i) $SS(\mathcal{K}) \subset \Lambda_\Phi \cup 0_{M \times M \times I};$
(ii) $\mathcal{K}|_{t=0} = \mathbf{k}_\Delta$, *where Δ is the diagonal of $M \times M$.*

Moreover, $\mathcal{K}$ is unique up to isomorphism in $\mathcal{D}(\mathbf{k}_{M \times M \times I})$. This $\mathcal{K}$ is called the sheaf quantization of Φ (and if necessary, one denotes it by $\mathcal{K}_\Phi$ to emphasize the dependence of Φ).

Remark 4.1 (1) The geometric constraint, item (i) in Theorem 4.1, is important, otherwise we can simply take $\mathcal{K} \equiv \mathbf{k}_\Delta$. (2) For most symplectic geometers, the time parameter usually ranges in $I = [0, 1]$, a closed interval. We could state Theorem 4.1 in terms of this closed interval, but then we would face the issue of defining correctly the singular support on a manifold with boundary. With an open interval I in Theorem 4.1, which is consistent with the statement of the main result in [26], the sheaf quantization $\mathcal{K}$ more precisely lies in $\mathcal{D}^{\mathrm{lb}}(\mathbf{k}_{M \times M \times I})$, where $\mathcal{D}^{\mathrm{lb}}$ means *locally bounded* derived category (see Example 3.11 in [26] for its necessity). Here, locally bounded means that for every relatively compact open subset $U \subset M \times M \times I$, the complex $\mathcal{K}|_U$ is bounded, i.e., $\mathcal{K}|_U \in \mathcal{D}^{\mathrm{b}}(\mathbf{k}_{M \times M \times I})$. Note that $\mathcal{D}^{\mathrm{lb}}(\mathbf{k}_{M \times M \times I})$ is strictly larger than $\mathcal{D}^{\mathrm{b}}(\mathbf{k}_{M \times M \times I})$.

Using the convolution operator, we can, for any $\mathcal{F} \in \mathcal{D}(\mathbf{k}_M)$, consider $\mathcal{K} \circ \mathcal{F} \in \mathcal{D}(\mathbf{k}_{M \times \mathbb{R}})$. Denote by $\dot{S}S(\mathcal{F})$ the part of $SS(\mathcal{F})$ away from the zero-section. By the geometric meaning of the convolution operator,

$$SS(\mathcal{K} \circ \mathcal{F}) \subset SS(\mathcal{K}) \circ SS(\mathcal{F})$$

$$\subset (\Lambda_\Phi \cup 0_{M \times M \times I}) \circ SS(\mathcal{F})$$

$$\subset (\Lambda_\Phi \cup 0_{M \times M \times I}) \circ (\dot{S}S(\mathcal{F}) \cup (\mathrm{supp}(\mathcal{F}) \times \{0\}))$$

$$\subset (\Lambda_\Phi \circ \dot{S}S(\mathcal{F})) \cup 0_{M \times I}$$

$$= \{(\phi_t(\dot{S}S(\mathcal{F})), (t, -F_t(\phi_t(\dot{S}S(\mathcal{F})))))\} \cup 0_{M \times I}.$$

In particular, restricted at any $t \in I$, $SS((\mathcal{K} \circ \mathcal{F})|_t) = SS(\mathcal{K}|_t \circ \mathcal{F}) \subset \phi_t(\dot{S}S(\mathcal{F})) \cup 0_M$. In other words, convolution with $\mathcal{K}|_t$ corresponds to "moved by ϕ_t". This leads to an important property of the restricted version of Tamarkin category (see Lemma 4.3).

---

[1] Here, "compactly supported" means $\mathrm{supp}(\phi_t)/\mathbb{R}_+$ is compact.

The following corollary of Theorem 4.1 proves very useful.

Corollary 4.1 *Suppose that M is a compact manifold and Φ is a compactly supported homogeneous Hamiltonian isotopy on $\dot{T}^*M$. Then for any $t \in I$ and any $\mathcal{F} \in \mathcal{D}(\mathbf{k}_M)$,*

$$H^*(M; (\mathcal{K} \circ \mathcal{F})|_t) = H^*(M; \mathcal{F}).$$

Example 4.1 A very special case of Corollary 4.1 is obtained when $\mathcal{F} = \mathbf{k}_N$, where N is a compact submanifold of M. Then

$$H^*(M; (\mathcal{K} \circ \mathbf{k}_N)|_t) = H^*(M; \mathbf{k}_N) \, (= H^*(N; \mathbf{k}) \neq 0).$$

A direct application of this corollary is the following non-displaceability result that is emphasized in [26],

Corollary 4.2 *Suppose that, $\Phi = \{\phi_t\}_{t \in I}$ is a homogeneous Hamiltonian isotopy and $\psi : M \to \mathbb{R}$ is a function such that $d\psi(x) \neq 0$ for any $x \in M$. Let $\mathcal{F} \in \mathcal{D}(\mathbf{k}_M)$. If $H^*(M; \mathcal{F}) \neq 0$, then for any $t \in I$,*

$$\phi_t(\dot{S}S(\mathcal{F})) \cap \mathrm{graph}(d\psi) \neq \emptyset.$$

Proof By Corollary 4.1, for any $t \in I$, $H^*(M; (\mathcal{K} \circ \mathcal{F})|_t) = H^*(M; \mathcal{F}) \neq 0$. Then by the microlocal Morse lemma (see Theorem 2.10), $SS((\mathcal{K} \circ \mathcal{F})|_t) \cap \mathrm{graph}(d\psi) \neq \emptyset$. On the other hand, as we have seen earlier,

$$SS((\mathcal{K} \circ \mathcal{F})|_t) \subset \phi_t(\dot{S}S(\mathcal{F})) \cup 0_M.$$

Then the hypothesis on ψ implies that $\phi_t(\dot{S}S(\mathcal{F})) \cap \mathrm{graph}(d\psi) \neq \emptyset$.   □

Example 4.2 Continued from Example 4.1; let $N \subset M$ be a compact submanifold of M; then Corollary 4.2 says that

$$\phi_t(\dot{\nu}^*N) \cap \mathrm{graph}(d\phi) \neq \emptyset. \tag{4.1}$$

Actually, N can be much worse than a submanifold (for instance, a submanifold with corners or singularities), yet $SS(\mathbf{k}_N)$ is still computable. It is a promising research direction to explore whether some (clever) choice of N and ψ can yield new rigidity result in terms of intersections in symplectic geometry.

Exercise 4.1 (see Theorem 4.16 in [26]). Use Corollary 4.2 (more precisely, Example 4.2) to prove the (Lagrangian) Arnold conjecture: Suppose that $\Phi = \{\phi_t\}_{t \in I} : I \times T^*M \to T^*M$ is a compactly supported Hamiltonian isotopy. Then

$$\phi_t(0_M) \cap 0_M \neq \emptyset.$$

In other words, the zero-section 0_M is non-displaceable.

Remark 4.2 With some further work based on Theorem 2.11, one is able to obtain a lower bound of the cardinality of the intersection in (4.1) (under a certain transversality assumption). In other words, Exercise 4.1 yields a finer formulation of the Arnold conjecture: $\#(\phi_t(0_M) \cap 0_M) \geq \sum \dim H^j(M; \mathbf{k})$.

4.1.2 Proof of a Simplified Sheaf Quantization

Here is an "oversimplified" version of Theorem 4.1 by considering only the time-1 map $\phi = \phi_1$ instead of the entire isotopy $\{\phi_t\}_{t \in I}$.

Theorem 4.2 *Let M be a manifold. Suppose that $\phi : \dot{T}^*M \to \dot{T}^*M$ is a homogeneous and compactly supported Hamiltonian diffeomorphism. Then there exists an element $\mathcal{K} \in \mathcal{D}(\mathbf{k}_{M \times M})$ such that*

$$SS(\mathcal{K}) \subset \mathrm{graph}(\phi)^{\mathrm{a}} \cup 0_{M \times M}$$

where "a" means changing the sign of the co-vector part. Moreover, when $\phi = \mathbb{1}$, we have $\mathcal{K} = \mathbf{k}_\Delta$.

One can view Theorem 4.2 as a corollary of Theorem 4.1. The graph $\mathrm{graph}(\phi)^{\mathrm{a}}$ is just Λ_Φ restricted at $t = 1$ and $\mathcal{K} = \mathcal{K}_\Phi|_{t=1}$. In this subsection, we give a proof of Theorem 4.2[2], which reveals the secret of the proof of the existence part of Theorem 4.1. However, "oversimplified" means that we cannot prove the uniqueness part of Theorem 4.1 without referring to the isotopy version. In fact, the key to proving this uniqueness is the same as the one used to prove Corollary 4.1. We postpone this to the next subsection. The construction of $\mathcal{K}$ in Theorem 4.2 needs to start from somewhere, and the following example, displayed in [26] for the first time, is a good choice.

Example 4.3 Choose a complete Riemannian metric on a manifold M with its injectivity radius at least ϵ_0. Denote by $g_t : \dot{T}^*M \to \dot{T}^*M$ the homogeneous geodesic flow. Let d denote the distance function. Consider

$$U = \{(x, y) \in M \times M \mid d(x, y) < \epsilon\} \tag{4.2}$$

for some $0 \leq \epsilon \ll \epsilon_0$, $\partial \bar{U} = \{d(x, y) = \epsilon\}$. If Δ is the diagonal of $M \times M$, and N is any closed subset of $M \times M$, then:

(1) $\mathbf{k}_{\bar{U}} \circ \mathbf{k}_U[n] = \mathbf{k}_U[n] \circ \mathbf{k}_{\bar{U}} = \mathbf{k}_\Delta$.
(2) $\mathbf{k}_\Delta \circ \mathbf{k}_\Delta = \mathbf{k}_\Delta$.
(3) $\mathbf{k}_\Delta \circ \mathbf{k}_N = \mathbf{k}_N \circ \mathbf{k}_\Delta = \mathbf{k}_N$.

---

[2]This version is based on Leonid Polterovich's lecture at the Kazhdan seminar held at the Hebrew University of Jerusalem in the Fall of 2017.

Remark 4.3 The only item difficult to prove is the first equality in (1) (which we call "dual sheaf identity"). For the reader's convenience, we will provide an elementary proof of it in Subsect. 4.1.4. This equality is used to confirm the second conclusion in Theorem 4.2.

Exercise 4.2 Check that

$$v_+^*(\partial U) = \text{graph}(g_{-\epsilon})^a$$

and

$$v_-^*(\partial U) = \text{graph}(g_\epsilon)^a.$$

Hint: do it first in the Euclidean case.

Then, by Example 2.16, we know $SS(\mathbf{k}_{\bar{U}}) = \text{graph}(g_\epsilon)^a \cup 0_{\bar{U}}$ and $SS(\mathbf{k}_U) = SS(\mathbf{k}_U[n]) = \text{graph}(g_{-\epsilon})^a \cup 0_{\bar{U}}$. Now, we are ready to give the proof.

Proof (of Theorem 4.2) Choose a sufficiently large $N \in \mathbb{N}$ and decompose

$$\phi = \prod_{i=1}^{N} \phi^{(i)} \quad \text{where for each } i \ \|\phi^{(i)}\|_{C^1} \ll 1.$$

Note that $\phi = \prod_{i=1}^{N} g_{-\epsilon} \circ (g_\epsilon \circ \phi^{(i)})$, where "$\circ$" is the composition of diffeomorphisms. Since $g_\epsilon \circ \phi^{(i)}$ is a small perturbation of g_ϵ, for each $i \in \{1, \dots, N\}$, there exists a small perturbation of U, denoted $U(i)$, such that

$$\text{graph}(g_\epsilon \circ \phi^{(i)})^a = v_-^*(\partial \overline{U(i)}). \quad \textbf{(Exercise)} \tag{4.3}$$

Consider the sheaf

$$\mathcal{K} := \prod_{i=N}^{1} (\mathbf{k}_{\overline{U(i)}} \circ \mathbf{k}_U[n]) \in \mathcal{D}(\mathbf{k}_{M \times M}). \tag{4.4}$$

By the functorial properties of singular supports,

$$SS(\mathcal{K}) \subset \prod_{i=N}^{1} \left(\text{graph}(g_\epsilon \circ \phi^{(i)})^a \circ \text{graph}(g_{-\epsilon})^a \right) \cup 0_{M \times M}$$

$$= \text{graph}(\phi)^a \cup 0_{M \times M}.$$

Finally, when $\phi = \mathbb{1}$, we can simply take $N = 1$ and $U(i) = U$, and there is no perturbation at all. Hence, by our construction and (i) in Example 4.3,

$$\mathcal{K} = \mathbf{k}_{\bar{U}} \circ \mathbf{k}_U[n] = \mathbf{k}_\Delta.$$

This means that this $\mathcal{K}$ is the sought-for sheaf, which completes the proof. □

Remark 4.4 It is easy to modify the proof above to obtain the existence of a (sought-for) sheaf $\mathcal{K} \in \mathcal{D}(\mathbf{k}_{M \times M \times I})$, simply by starting from $U_t = \{(x, y, t) \in M \times M \times I \mid d(x, y) < t\}$ where t is also viewed as a variable.

4.1.3 Constraints Coming from Singular Supports

In this subsection, we prove two propositions: one implies the uniqueness part in Theorem 4.1 and the other implies Corollary 4.1. Both of them follow from Lemma 2.3, which shows that constraints of singular supports can impose strong restrictions on the behavior of sheaves. The maps in these propositions are depicted in the following diagram:

$$
\begin{array}{ccc}
X \times I & \xrightarrow{\ q\ } & I. \\
{\scriptstyle p}\big\downarrow & & \\
X & &
\end{array}
$$

Proposition 4.1 ((ii) in Proposition 5.4.5 in [32]) *If $\mathcal{F} \in \mathcal{D}(\mathbf{k}_{X \times I})$ satisfies $SS(\mathcal{F}) \subset T^*X \times 0_I$, then $\mathcal{F} \simeq p^{-1}Rp_*\mathcal{F}$, where $p : X \times I \to X$ is the canonical projection.*

Proposition 4.2 (Deformation of coefficient) *If $\mathcal{F} \in \mathcal{D}(\mathbf{k}_{X \times I})$ satisfies $SS(\mathcal{F}) \cap (0_X \times T^*I) \subset 0_{X \times I}$, then $H^*(X; \mathcal{F}|_s) = H^*(X; \mathcal{F}|_t)$.*

Before we prove them, let us see how they imply what we promised.

Proof (of the uniqueness in Theorem 4.1) For $\mathcal{F} := \mathcal{K}_{\Phi^{-1}} \circ_M \mathcal{K}_\Phi \in \mathcal{D}(\mathbf{k}_{M \times M \times I})$, it is easy to check that $SS(\mathcal{F}) \subset T^*(M \times M) \times 0_I$. Denote $\iota_t : M \times M \times \{t\} \to M \times M \times I$. By Proposition 4.1 (where $X = M \times M$), $\mathcal{F} = p^{-1}Rp_*\mathcal{F}$. Therefore,

$$
\begin{aligned}
\mathcal{F}|_t &= \iota_t^{-1}(p^{-1}Rp_*\mathcal{F}) \\
&= (p \cdot \iota_t)^{-1}Rp_*\mathcal{F} \\
&= Rp_*\mathcal{F} \\
&= (p \cdot \iota_0)^{-1}Rp_*\mathcal{F} \\
&= \iota_0^{-1}(p^{-1}Rp_*\mathcal{F}) \\
&= \mathcal{F}|_{t=0} = \mathbf{k}_\Delta.
\end{aligned}
$$

In particular, $\mathcal{F}|_t$ is independent of $t \in I$. So $\mathcal{K}_{\Phi^{-1}} \circ_M \mathcal{K}_\Phi(= \mathcal{F}) = p^{-1}Rp_*\mathcal{F} = \mathbf{k}_{\Delta \times I}$. Similarly, $\mathcal{K}_\Phi \circ_M \mathcal{K}_{\Phi^{-1}} = \mathbf{k}_{\Delta \times I}$. Hence, if $\mathcal{K}_1$ and $\mathcal{K}_2$ are both sheaf quantizations of Φ, then

$$\mathcal{K}_1 \simeq \mathcal{K}_1 \circ_M (\mathcal{K}_{\Phi^{-1}} \circ_M \mathcal{K}_2)$$

$$\simeq (\mathcal{K}_1 \circ_M \mathcal{K}_{\Phi^{-1}}) \circ_M \mathcal{K}_2$$

$$\simeq \mathcal{K}_2$$

The proof is complete.   □

Remark 4.5 The original proof of Proposition 4.1 given in [26] introduces an "inverse quantization", denoted by $\mathcal{K}_\Phi^{-1}$ (and more generally, one can define an inverse/dual sheaf $\mathcal{K}^{-1}$ of a given sheaf $\mathcal{K}$ on any manifold of finite dimension). Here, the name used for $\mathcal{K}_\Phi^{-1}$ is justified by the property that

$$\mathcal{K}_\Phi^{-1} \circ_M \mathcal{K}_\Phi \simeq \mathcal{K}_\Phi \circ_M \mathcal{K}_\Phi^{-1} \simeq \mathbf{k}_{\Delta \times I}.$$

By the uniqueness in Theorem 4.1, $\mathcal{K}_\Phi^{-1} \simeq \mathcal{K}_{\Phi^{-1}}$. Moreover, the association of sheaf quantizations to Hamiltonian isotopies admits a group structure summarized in the following table

dynamics	sheaf	geometry
Φ	$\mathcal{K}_\Phi$	Λ_Φ
$\Phi \circ \Psi$	$\mathcal{K}_\Phi \circ_M \mathcal{K}_\Psi$	$\Lambda_\Phi \circ \vert_I \Lambda_\Psi$
Φ^{-1}	$\mathcal{K}_{\Phi^{-1}}$	$\Lambda_{\Phi^{-1}}$

where $\Lambda_\Phi \circ \vert_I \Lambda_\Psi$ is a time-dependent version of Lagrangian correspondence (cf. Remark 3.12), i.e., Lagrangian correspondence on the M-component but summation of co-vectors on the t-component. From sheaf to geometry, this transformation is confirmed by the following relation (see (1.15) in [26]), which can be thought as a time-dependent generalization of (3.5):

$$SS(\mathcal{K}_\Phi \circ_M \mathcal{K}_\Psi) \subset SS(\mathcal{K}_\Phi) \circ \vert_I SS(\mathcal{K}_\Psi) \subset \Lambda_\Phi \circ \vert_I \Lambda_\Psi.$$

Proof *(of Corollary 4.1)* View the given $\mathcal{F}$ as $(\mathcal{K} \circ \mathcal{F})\vert_{t=0} (= \mathbf{k}_\Delta \circ \mathcal{F}) \in \mathcal{D}(\mathbf{k}_M)$. We have an I-parametrized family of sheaves, that is, $\mathcal{G} = \mathcal{K} \circ \mathcal{F} \in \mathcal{D}(\mathbf{k}_{M \times I})$ with $\mathcal{G}\vert_{t=0} = \mathcal{F}$. We have seen that

$$SS(\mathcal{G}) \subset \{(\phi_t(\dot{SS}(\mathcal{F})), (t, -F_t(\phi_t(\dot{SS}(\mathcal{F})))))\} \cup 0_{M \times I}.$$

Since ϕ_t acts on $\dot{T}^*M$, intersection with $0_M \times T^*I$ only results in the zero-section part. Proposition 4.2 with $X = M$ yields the desired conclusion once we take $s = 0$ and any $t \in I$.   □

The rest of this subsection will be devoted to the proof of Proposition 4.1 and 4.2.

Proof *(of Proposition 4.1)* Since p^{-1} and Rp_* are adjoint to each other, there exists a well-defined map $p^{-1}Rp_*\mathcal{F} \to \mathcal{F}$. Now we only need to check the identity on the level of stalks. First,

$$(p^{-1}R^j p_*\mathcal{F})_{(x,t)} = (R^j p_*\mathcal{F})_x = H^j(p^{-1}(x); \mathcal{F}|_{p^{-1}(x)}).$$

Then a key observation is that $\mathcal{F}|_{p^{-1}(x)}$ is a locally constant sheaf by our assumption $SS(\mathcal{F}|_{p^{-1}(x)}) \subset 0_I$, which implies that it is actually constant since $p^{-1}(x) = I$ is contractible. Since $(x, t) \in p^{-1}(x)$, if we denote $\mathcal{F}_{(x,t)} := V[d]$ where $V = \mathbf{k}^m$ for some dimension m and d is the degree, then the only non-zero degree of $p^{-1}Rp_*\mathcal{F}$ is

$$H^j(p^{-1}(x); \mathcal{F}|_{p^{-1}(x)}) = H^0(I; V(I)[d]) = V[d],$$

where $V(I)$ denotes the constant sheaf over I with its stalk being the $\mathbf{k}$-module V. Therefore, one gets that $(p^{-1}R^0 p_*\mathcal{F})_{(x,t)} = \mathcal{F}_{(x,t)}$. This implies that, as two elements in derived category, $p^{-1}Rp_*\mathcal{F}$ is quasi-isomorphic to $\mathcal{F}$. □

Proof *(of Proposition 4.2)* Applying the pushforward formula for singular support (see Proposition 2.6), one gets

$$SS(Rq_*\mathcal{F}) \subset \{(t, \tau) \in T^*I | \exists (x, \xi, t, \tau) \in SS(\mathcal{F}) \text{ s.t. } q(x,t)=t \text{ and } q^*(\tau)=(\xi, \tau)\}.$$

Since $q^*(\tau) = (0, \tau)$, $\xi = 0$. Hence, by our assumption, $\tau = 0$. Therefore, $SS(Rq_*\mathcal{F}) \subset 0_I$ and then $Rq_*\mathcal{F}$ is a constant sheaf. It implies that

$$R\Gamma(M, \mathcal{F}|_s) = (Rq_*\mathcal{F})_s = (Rq_*\mathcal{F})_t = R\Gamma(M, \mathcal{F}|_t),$$

as we needed to show. □

So far we have seen how to associate a unique sheaf $\mathcal{K}_\Phi$ to a Hamiltonian isotopy (1-parameter family of Hamiltonian diffeomorphisms). It can be checked that the same argument can be invoked to associate a unique sheaf $\mathcal{K}_\Theta \in \mathcal{D}(\mathbf{k}_{M \times M \times I \times I})$ to a 2-*parameter* family of Hamiltonian diffeomorphisms $\Theta = \{\theta_{(t,s)}\}_{(t,s) \in I^2} : T^*M \times I \times I \to T^*M$, where each $\theta_{t,s}$ is a Hamiltonian diffeomorphism and $\theta_{(0,0)} = \mathbb{1}$, such that (i) $SS(\mathcal{K}_\Theta) \subset \Lambda_\Theta \cup 0_{M \times M \times I \times I}$, where

$$\Lambda_\Theta = \{((x, \xi), -\theta_{s,t}(x, \xi), (s, -H_{t,s}(\theta_{t,s}(x, \xi))), (t, -F_{(t,s)}(\theta_{t,s}(x, \xi))))\} \tag{4.5}$$

and $H_{t,s}$ and $F_{t,s}$ are Hamiltonian functions corresponding to the vector fields in s-direction and t-direction respectively; (ii) $\mathcal{K}_\Theta|_{(0,0)} = \mathbf{k}_\Delta$.

Let us consider a special 2-parameter family of Hamiltonian diffeomorphisms. Suppose Φ and Ψ are Hamiltonian isotopies with fixed end points, equal to $\mathbb{1}$ when $t = 0$ and equal to some ϕ when $t = 1$. Moreover, we require that Φ and Ψ are homotopic through Hamiltonian isotopies with fixed endpoints. This homotopy is parametrized by s, and we get a 2-parameter family of Hamiltonian isotopies Θ. In particular, $H_{1,s} \equiv 0$ in (4.5). Now, we can state the following interesting proposition asserting that the sheaf quantization is in fact unique up to any homotopy through Hamiltonian isotopics.

Proposition 4.3 *Suppose that Φ and Ψ are Hamiltonian isotopies, homotopic with fixed end points. Then $\mathcal{K}_\Phi|_{t=1} \simeq \mathcal{K}_\Psi|_{t=1}$.*

Proof It's easy to check that $\mathcal{K}_\Theta|_{s=0}$ is a sheaf quantization of Φ (satisfying (i) and (ii) in Theorem 4.1). By the uniqueness in Theorem 4.1, $\mathcal{K}_\Theta|_{s=0} \simeq \mathcal{K}_\Phi$. Similarly, $\mathcal{K}_\Theta|_{s=1} \simeq \mathcal{K}_\Psi$. Therefore,

$$\mathcal{K}_\Phi|_{t=1} = \mathcal{K}_\Theta|_{s=0,t=1} = (\mathcal{K}_\Theta|_{t=1})|_{s=0}$$

and

$$\mathcal{K}_\Psi|_{t=1} = \mathcal{K}_\Theta|_{s=1,t=1} = (\mathcal{K}_\Theta|_{t=1})|_{s=1}.$$

Consider the sheaf $\mathcal{F} := \mathcal{K}_\Theta|_{t=1} \in \mathcal{D}(\mathbf{k}_{M \times I})$ where I is the range of the parameter s. Since $H_{1,s} \equiv 0$ by our assumption, $SS(\mathcal{F}) \subset T^*M \times 0_I$. By the same argument as in the proof of the uniqueness in Theorem 4.1, $\mathcal{F}|_s \simeq Rp_*\mathcal{F}$, where $p : M \times I \to M$ is the canonical projection, in particular, independent of $s \in I$. Therefore, $\mathcal{F}|_{s=0} \simeq \mathcal{F}|_{s=1}$, as claimed. $\square$

4.1.4 Proof of "Dual Sheaf Identity"

This is a technical subsection[3] devoted to proving (1) in Example 4.3, that is,

$$\mathbf{k}_{\bar{U}} \circ \mathbf{k}_U[n] = \mathbf{k}_U[n] \circ \mathbf{k}_{\bar{U}} = \mathbf{k}_\Delta.$$

For brevity, we will only give a proof of an easy version. First of all, it is very helpful to keep tracking the position of each factor in the product space, so we will denote $\mathbb{R}$ by $\mathbb{R}_x$ (or $\mathbb{R}_y$ and so on). For a fixed $\epsilon > 0$, simply denote

$$U_{xy} := \left\{ (x, y) \in \mathbb{R}_x \times \mathbb{R}_y \mid |x - y| < \epsilon \right\}$$

and

$$\overline{U_{yz}} := \left\{ (y, z) \in \mathbb{R}_y \times \mathbb{R}_z \mid |y - z| \leq \epsilon \right\}.$$

---

[3]This is based on conversations with Leonid Polterovich and Yakov Varshavsky.

Then we have the following 1-dimensional Euclidean space version of the "dual sheaf identity".

Proposition 4.4 *Denote the diagonal by* $\Delta_{xz} = \{(x, z) \in \mathbb{R}_x \times \mathbb{R}_z \mid x = z\}$. *We have the identity*

$$\mathbf{k}_{U_{xy}} \circ \mathbf{k}_{\overline{U_{yz}}} \simeq \mathbf{k}_{\Delta_{xz}}[-1],$$

where "$\simeq$" means quasi-isomorphic.

The following lemma will be used frequently.

Lemma 4.1 *Suppose that S is either a closed or an open subset of X. Let $f : Y \to X$ be a continuous map. Then $f^{-1}(\mathbf{k}_S) \simeq \mathbf{k}_{f^{-1}(S)}$.*

Proof This essentially follows from the base change formula. When S is closed, $\mathbf{k}_S = i_*(\mathbf{k}(S))$ where $\mathbf{k}(S)$ is the constant sheaf over S and $i : S \to X$ is simply the inclusion map; when S is open, $\mathbf{k}_S = i_!(\mathbf{k}(S))$. Now, assume that S is closed. Consider the cartesian square

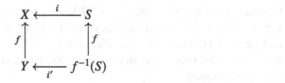

Starting from $\mathbf{k}(S)$, we have

$$f^{-1}(\mathbf{k}_S) = f^{-1}i_*(\mathbf{k}(S)) \simeq i'_* f^{-1}(\mathbf{k}(S)) = i'_* \mathbf{k}(f^{-1}(S)) = \mathbf{k}_{f^{-1}(S)}.$$

The third equality comes from rewriting the constant sheaf $\mathbf{k}(S) = a^{-1}\mathbf{k}$ for map $a : S \to \{\text{pt}\}$ and then $f^{-1}(\mathbf{k}(S)) = f^{-1}(a^{-1}\mathbf{k}) = (a \circ f)^{-1}(\mathbf{k})$. Note that usually the base change formula works for pushforward with compact support. Since here S is closed, $i_* = i_!$. The same argument works for S open.   □

The proof of Proposition 4.4 is quite complicated. First, preparations.

(a) **(composition formula)** Recall the composition formula of two sheaves,

$$\mathbf{k}_{U_{xy}} \circ \mathbf{k}_{\overline{U_{yz}}} = Rq_{xz!}(q_{xy}^{-1}\mathbf{k}_{U_{xy}} \otimes q_{yz}^{-1}\mathbf{k}_{\overline{U_{yz}}}),$$

where $q_{xy} : \mathbb{R}_x \times \mathbb{R}_y \times \mathbb{R}_z \to \mathbb{R}_x \times \mathbb{R}_y$ is the projection, and similarly for q_{yz} and q_{xz}. Denote $\mathcal{F} := \mathbf{k}_{U_{xy}} \circ \mathbf{k}_{\overline{U_{yz}}} \in \mathcal{D}(\mathbf{k}_{\mathbb{R}_x \times \mathbb{R}_z})$. One can show that $\mathcal{F}$ is only supported on Δ_{xz} (**Exercise**), therefore it suffices to consider $\mathcal{F}|_{\Delta_{xz}}$. Denote the diagonal inclusion by $i : \Delta_{xz} \to \mathbb{R}_x \times \mathbb{R}_z$; then $\mathcal{F}|_{\Delta_{xz}} = i^{-1}(\mathcal{F})$. Consider the cartesian square

$$\mathbb{R}_x \times \mathbb{R}_y \times \mathbb{R}_z \xleftarrow{\quad i' \quad} \Delta_{xz} \times \mathbb{R}_y$$

$$\left\downarrow q_{xz} \qquad\qquad\qquad\qquad \downarrow p\right.$$

$$\mathbb{R}_x \times \mathbb{R}_z \xleftarrow{\quad i \quad} \Delta_{xz} \qquad\qquad (4.6)$$

The base change formula tells us that

$$
\begin{aligned}
\mathcal{F}|_{\Delta_{xz}} = i^{-1}\mathcal{F} &= i^{-1} Rq_{xz!}(q_{xy}^{-1}\mathbf{k}_{U_{xy}} \otimes q_{yz}^{-1}\mathbf{k}_{\overline{U_{yz}}}) \\
&= Rp_! i'^{-1}(q_{xy}^{-1}\mathbf{k}_{U_{xy}} \otimes q_{yz}^{-1}\mathbf{k}_{\overline{U_{yz}}}) \\
&= Rp_! \left((q_{xy} \circ i')^{-1}\mathbf{k}_{U_{xy}} \otimes (q_{yz} \circ i')^{-1}\mathbf{k}_{\overline{U_{yz}}} \right).
\end{aligned}
$$

Hence, by Lemma 4.1, we have that

$$
(q_{xy} \circ i')^{-1}\mathbf{k}_{U_{xy}} = \mathbf{k}_{(q_{xy}\circ i')^{-1}(U_{xy})} \quad \text{and} \quad (q_{yz} \circ i')^{-1}\mathbf{k}_{\overline{U_{yz}}} = \mathbf{k}_{(q_{yz}\circ i')^{-1}(\overline{U_{yz}})}.
$$

$$(4.7)$$

(b) **(coordinate change)** In order to carry out computations efficiently in the proof of Proposition 4.4, we will change coordinates. Introduce new coordinates (x, s, t) by

$$x = x, \quad s = y - x \quad \text{and} \quad t = y - z.$$

Then we can rewrite some subsets appearing earlier as follows.

$$
\begin{aligned}
(q_{xy} \circ i')^{-1}(U_{xy}) &= \{(x, y, z) \mid x = z, \ |x - y| < \epsilon\} \\
&= \{(x, s, t) \mid x \in \mathbb{R}_x, \ s = t, \ |s| < \epsilon\} \quad \text{(in new coordinates)} \\
&= \mathbb{R}_x \times \Delta_{st}^{I_s},
\end{aligned}
$$

where Δ_{st} is the diagonal of $\mathbb{R}_s \times \mathbb{R}_t$, $I = (-\epsilon, \epsilon)$ (where I_s denotes I in terms of the s-coordinate and similarly I_t for the t-coordinate) and $\Delta_{st}^{I_s} = \{(s, t) \in \Delta_{st} \mid s \in I\}$. Further,

$$
(q_{yz} \circ i')^{-1}(\overline{U_{yz}}) = \{(x, s, t) \mid x \in \mathbb{R}_x, \ s = t, \ |t| \le \epsilon\} = \mathbb{R}_x \times \Delta_{st}^{\overline{I_t}}.
$$

Since $s = t$, we identify Δ_{st} with $\mathbb{R}$ (denoted as $\mathbb{R}_\Delta$). Then $I_s = I$ and $\overline{I_t} = \overline{I}$. With the projection $\pi_\Delta : \mathbb{R}_x \times \mathbb{R}_\Delta \to \mathbb{R}_\Delta$, Lemma 4.1 again yields

$$
(q_{xy} \circ i')^{-1}\mathbf{k}_{U_{xy}} = \pi_\Delta^{-1}\mathbf{k}_I = \mathbf{k}_{\mathbb{R}_x \times I} \qquad (4.8)
$$

and

$$(q_{yz} \circ i')^{-1} \mathbf{k}_{\overline{U_{yz}}} = \pi_\Delta^{-1} \mathbf{k}_{\bar{I}} = \mathbf{k}_{\mathbb{R}_x \times \bar{I}}. \tag{4.9}$$

Moreover, the projection p in square (4.6) can be identified with the projection $q : \mathbb{R}_x \times \Delta_{st}(= \mathbb{R}_x \times \mathbb{R}_\Delta) \to \mathbb{R}_x$, via

$$q(\{(x, s, t) \mid s = t\}) = \{x\}.$$

All together we can rewrite

$$\mathcal{F}|_{\Delta_{xz}} = Rq_! \left(\pi_\Delta^{-1} \mathbf{k}_I \otimes \pi_\Delta^{-1} \mathbf{k}_{\bar{I}} \right) = Rq_!(\pi_x^{-1} \mathbf{k}_{\mathbb{R}_x} \otimes \pi_\Delta^{-1} \mathbf{k}_I), \tag{4.10}$$

with the projection $\pi_x : \mathbb{R}_x \times \mathbb{R}_\Delta \to \mathbb{R}_x$ and $\mathbf{k}_I \otimes \mathbf{k}_{\bar{I}} = \mathbf{k}_I$. In fact, $\pi_x = q$.

(c) **(Künneth formula)** In general,

$$H^*(M \times N, p_M^{-1}\mathcal{F} \otimes p_N^{-1}\mathcal{G}) = \bigoplus_{i+j=*} H^i(M, \mathcal{F}) \otimes H^j(N, \mathcal{G}),$$

where $p_M : M \times N$ is the projection onto the M-component, and similarly for p_N.

Now we are ready.

Proof *(of Proposition 4.4)* We will show that $\mathcal{F}|_{\Delta_{xz}}$ satisfies the following property that, for any contractible $U \subset \Delta_{xz}$, its cohomologies

$$h^*(\mathcal{F}|_{\Delta_{xz}}(U)) = \begin{cases} \mathbf{k}, & \text{for } * = 1, \\ 0, & \text{otherwise}, \end{cases} \tag{4.11}$$

i.e., viewing $\mathcal{F}|_{\Delta_{xz}}$ as a complex of sheaves, its cohomologies are not zero only in degree $* = 1$. Moreover, this cohomology in degree $* = 1$, **as a sheaf**, is a constant sheaf over Δ_{xz}. Therefore $\mathcal{F}_{\Delta_{xz}}$ is quasi-isomorphic to $\mathbf{k}_{\Delta_{xz}}[-1]^4$, as we needed to show.

Now, let us prove (4.11). Each open subset $U \subset \Delta_{xz}$ can be identified with $U_x \subset \mathbb{R}_x$, where U_x is the projection of U on $\mathbb{R}_x$. Note that

$$q^{-1}(U_x) = U_x \times \mathbb{R}_\Delta.$$

---

[4]This comes from the following standard fact concerning the derived category: Let $\mathcal{A}$ be an abelian category. If $\mathcal{G}_\bullet \in \mathcal{D}(\mathcal{A})$ has non-zero cohomology only in degree k, then $\mathcal{G}_\bullet$ is quasi-isomorphic to the single term complex $(0 \to h^k(\mathcal{G}_\bullet) \to 0)$ **(Exercise)**.

By (b) and (c) above, $Rq_! \left(\pi_x^{-1} \mathbf{k}_{\mathbb{R}_x} \otimes \pi_\Delta^{-1} \mathbf{k}_I \right) (U_x)$ computes the cohomology (of a complex of $\mathbf{k}$-modules),

$$H_{\mathrm{cv}}^*(U_x \times \mathbb{R}_\Delta, \pi_x^{-1} \mathbf{k}_{\mathbb{R}_x} \otimes \pi_\Delta^{-1} \mathbf{k}_I) = \bigoplus_{i+j=*} H^i(U_x, \mathbf{k}_{\mathbb{R}_x}) \otimes H_c^j(\mathbb{R}_\Delta, \mathbf{k}_I),$$

where H_{cv}^* means the *vertically* compactly supported cohomology, and "vertically" means the direction along fibers. Here, the fiber of q is $\mathbb{R}_\Delta$ (therefore, we compute H_c^* for the second factor). Obviously, the first part

$$H^i(U_x, \mathbf{k}_{\mathbb{R}_x}) = \begin{cases} \mathbf{k}, & \text{for } i = 0, \\ 0, & \text{otherwise.} \end{cases} \tag{4.12}$$

Now, we claim that

$$H_c^j(\mathbb{R}_\Delta, \mathbf{k}_I) = \begin{cases} \mathbf{k}, & \text{for } j = 1, \\ 0, & \text{otherwise.} \end{cases} \tag{4.13}$$

In fact, consider

$$I \xrightarrow{\ i\ } \mathbb{R}(= \mathbb{R}_\Delta) \xrightarrow{\ a\ } \{\mathrm{pt}\}.$$

We have the following computation:

$$
\begin{aligned}
R\Gamma_c(\mathbb{R}, \mathbf{k}_I) &= R\Gamma_c(\mathbb{R}, i_! \mathbf{k}(I)) && \mathbf{k}(I) \text{ is the constant sheaf over the } \mathbf{open} \ I \\
&= R\Gamma_c(\mathbb{R}, Ri_! \mathbf{k}(I)) && \text{because } i_! \text{ is exact here} \\
&= R(\Gamma_c(\mathbb{R}, \cdot) \circ i_!)(\mathbf{k}(I)) && \text{by the Grothendieck composition formula} \\
&= R(a_! \circ i_!)(\mathbf{k}(I)) && \text{by the definition of } \Gamma_c(\mathbb{R}, \cdot) \\
&= R((a \circ i)_!)(\mathbf{k}(I)) && \text{by the functorial property} \\
&= R\Gamma_c(I, \mathbf{k}) \Rightarrow (4.13) && \text{by the de Rham cohomology.}
\end{aligned}
$$

Finally, a degree counting using (4.12) and (4.13) gives the desired conclusion. $\square$

4.2 Stability with Respect to $d_{\mathcal{T}(M)}$

Recall that the sheaf quantization theorem in [26] or Theorem 4.1 in the previous section is stated in terms of homogeneous Hamiltonian isotopies. We next present a natural example (which is also the starting point of the construction of $\mathcal{K}_\Phi$ in Subsect. 4.1.2).

Example 4.4 (Homogeneous geodesic) Let $\Phi : I \times \dot{T}^*\mathbb{R}^n \to \dot{T}^*\mathbb{R}^n$ be the flow $(s, (x, \xi)) \mapsto (x - s\xi/|\xi|, \xi)$, which is generated by $H_s(x, \xi) = |\xi|$. Then

$$\mathcal{K}_\Phi = \mathbf{k}_{\{(s,x,y)\,|\,d(x,y)\leq s\}}.$$

Denote $Z := \{(s, x, y) \,|\, d(x, y) \leq s\}$. Note that $SS(\mathcal{K}_\Phi) = \nu_-^*(\partial Z)$ (the negative part of the conormal bundle).

Given a compactly supported Hamiltonian isotopy $\phi = \{\phi_s\}_{s\in I} : I \times T^*M \to T^*M$ generated by $h_s : I \times T^*M \to \mathbb{R}$, there exists a (Hamiltonian) lift $\Phi : I \times T^*_{\{\tau>0\}}(M \times \mathbb{R}) \to T^*_{\{\tau>0\}}(M \times \mathbb{R})$, defined by

$$\Phi(s, (m, \xi, t, \tau)) = (\tau \cdot \phi_s(m, \xi/\tau), t + (*), \tau)$$

for some function $(*)$. Note that this lifted Φ is generated by a homogeneous Hamiltonian function

$$H_s(m, \xi, t, \tau) = \tau \cdot h_s(m, \xi/\tau).$$

Formally, this lift/homogenization can be regarded as a trick to fit into the set-up of Theorem 4.1. On the other hand, in Appendix A.3 we give a dynamical explanation of this lift/homogenization.

Denote the sheaf quantization of the lift of the Hamiltonian isotopy ϕ simply by $\mathcal{K}(\phi) \in \mathcal{D}(\mathbf{k}_{I\times M\times\mathbb{R}\times M\times\mathbb{R}})$. From Example 3.7 we know that

$$\mathcal{K}(\phi)|_{s=1}\circ : \mathcal{D}(\mathbf{k}_{M\times\mathbb{R}}) \longrightarrow \mathcal{D}(\mathbf{k}_{M\times\mathbb{R}}).$$

In fact, we have a stronger result as follows,

Lemma 4.2 $\mathcal{K}(\phi)|_{s=1}\circ$ *is a well-defined morphism in the Tamarkin category* $\mathcal{T}(M)$.

Proof For any $\mathcal{F} \in \mathcal{T}(M)$,

$$(\mathcal{K}(\phi)|_{s=1} \circ \mathcal{F}) * \mathbf{k}_{M\times[0,\infty)} = \mathcal{K}(\phi)|_{s=1} \circ (\mathcal{F} * \mathbf{k}_{M\times[0,\infty)}) = \mathcal{K}(\phi)|_{s=1} \circ \mathcal{F},$$

by Theorem 3.1.   □

Exercise 4.3 Verify the first equality in the proof above. Hint: $\mathcal{F} * \mathbf{k}_{M\times[0,\infty)} = \mathcal{F} \circ_{\mathbb{R}} \delta^{-1}\mathbf{k}_{[0,\infty)}$ where $\delta : \mathbb{R}^2 \to \mathbb{R}$ by $(x, y) \to y - x$.

Lemma 4.3 *Let* $\mathcal{F} \in \mathcal{T}_A(M)$. *Then* $\mathcal{K}(\phi)|_{s=1} \circ \mathcal{F} \in \mathcal{T}_{\phi_1(A)}M$.

Proof It is easy to check that for a given subset $C \subset T^*_{\{\tau>0\}}(M \times \mathbb{R})$, $\Lambda_\Psi \circ C = \Psi(C)$ for any Hamiltonian diffeomorphism Ψ on $T^*_{\{\tau>0\}}(M \times \mathbb{R})$. Therefore, for $A \subset T^*M$ and $\rho : T^*_{\{\tau>0\}}(M \times \mathbb{R}) \to T^*M$ (the reduction map), one has

$$\Lambda_{\Phi_1} \circ \rho^{-1}(A) = \Phi_1(\rho^{-1}(A))$$
$$= \Phi_1(\{(m, \tau\xi, t, \tau) \mid (m, \xi) \in A, \tau > 0\})$$
$$= \{(\tau \cdot \phi_1(m, \xi), t + (*), \tau) \mid (m, \xi) \in A, \tau > 0\}$$
$$\subset \rho^{-1}(\phi_1(A)).$$

On the other hand, in view of the geometric meaning of sheaf composition (see the formula (3.5)),

$$SS(\mathcal{K}(\phi)|_{s=1} \circ \mathcal{F}) \subset SS(\mathcal{K}(\phi)|_{s=1}) \circ SS(\mathcal{F})$$
$$\subset \Lambda_{\Phi_1} \circ \rho^{-1}(A)$$
$$\subset \rho^{-1}(\phi_1(A)).$$

The computation above together with Lemma 4.2 yield the desired conclusion. $\quad\square$

It is worth pointing out that for any $\mathcal{F} \in \mathcal{T}(M)$, $\mathcal{K}(\phi) \circ \mathcal{F} \in \mathcal{D}(\mathbf{k}_{I \times M \times \mathbb{R}})$, and

$$SS(\mathcal{K}(\phi) \circ \mathcal{F}) \subset \{(s, -\tau \cdot h_s(m, \xi/\tau)), (\tau \cdot \phi_s(m, \xi/\tau), t + (*), \tau) \mid (m, \xi, t, \tau) \in SS(\mathcal{F})\}$$
$$\subset \{(t, s, \tau, -h_s(m, \xi/\tau) \cdot \tau) \mid (m, \xi) \in T^*M\} \times T^*M$$

In other words,

$$SS(\mathcal{K}(\phi) \circ \mathcal{F}) \subset \left\{ (t, s, \tau, \sigma) \,\middle|\, -\max_{(m,\xi)} h_s \cdot \tau \leq \sigma \leq -\min_{(m,\xi)} h_s \cdot \tau \right\} \times T^*M. \tag{4.14}$$

The relation (4.14) gives a (cone shape) constraint on the singular support, with slopes depending on s. This is similar to (3.15) or (3.16). This observation leads to the following key theorem (see Theorem 4.16 in [1]). Recall the definition (1.7) of the Hofer norm in Sect. 1.6.

Theorem 4.3 *For any $\mathcal{F} \in \mathcal{T}(M)$, $d_{\mathcal{T}(M)}(\mathcal{F}, \mathcal{K}(\phi)|_{s=1} \circ \mathcal{F}) \leq \|\phi\|_{\mathrm{Hofer}}$.*

Proof Suppose that ϕ is generated by $h_s : I \times M \to \mathbb{R}$. We aim to show that $\mathcal{F}$ and $\mathcal{K}(\phi)|_{s=1} \circ \mathcal{F}$ are $\|h_s\|_{\mathrm{Hofer}}$-interleaved. Divide $I = [0, 1]$ into n segments, by the points, $s_i = i/n$. By Exercise 3.8, $\mathcal{K}(\phi)|_{s=s_i} \circ \mathcal{F}$ and $\mathcal{K}(\phi)|_{s=s_{i+1}} \circ \mathcal{F}$ are

$$\frac{1}{n} \max_{s \in [s_i, s_{i+1}]} \left(\max_{(m,\xi)} h_s - \min_{(m,\xi)} h_s \right) \text{-interleaved.}$$

Consequently, $\mathcal{F}(= \mathcal{K}(\phi)|_{s=0} \circ \mathcal{F})$ and $\mathcal{K}(\phi)|_{s=1} \circ \mathcal{F}$ are

$$\sum_{i=0}^{n} \frac{1}{n} \max_{s \in [s_i, s_{i+1}]} \left(\max_{(m,\xi)} h_s - \min_{(m,\xi)} h_s \right) \text{-interleaved.}$$

Letting $n \to \infty$ and using the definition of a Riemann integral, we conclude that $\mathcal{F}(= \mathcal{K}(\phi)|_{s=0} \circ \mathcal{F})$ and $\mathcal{K}(\phi)|_{s=s_{i+1}} \circ \mathcal{F}$ are

$$\left(\int_0^1 \max_{(m,\xi)} h_s - \min_{(m,\xi)} h_s \, ds \right) \text{-interleaved,}$$

as required.   □

4.3   Energy-Capacity Inequality (Following Asano and Ike)

Recall that Definition 3.9 and Definition 3.10 have introduced the adjoint sheaf $\bar{\bar{\mathcal{F}}}$ and the internal hom $\mathcal{H}om^*$ in Tamarkin category $\mathcal{T}(M)$, both of which were used to prove the Separation Theorem. In this section, we will see that they can also be used to define a capacity of a domain $A \subset T^*M$.

Definition 4.2 (Capacity of a sheaf $\mathcal{F} \in \mathcal{T}(M)$) For a sheaf $\mathcal{F} \in \mathcal{T}(M)$, we define the capacity of $\mathcal{F}$ by

$$c(\mathcal{F}) = \inf\{c > 0 \mid R\pi_* \mathcal{H}om^*(\mathcal{F}, \mathcal{F}) \text{ is } c\text{-torsion}\}.$$

Remark 4.6 Note that we can also define

$$c'(\mathcal{F}) = \inf\{c > 0 \mid R\text{Hom}(\mathcal{F}, \mathcal{F}) \to R\text{Hom}(\mathcal{F}, T_{c*}\mathcal{F}) \text{ is } 0\}$$
$$= \inf\{c > 0 \mid \tau_c(F) = 0\}.$$

Since $R\text{Hom}(\mathcal{F}, T_{c*}\mathcal{F}) = R\text{Hom}(\mathbf{k}_{[0,\infty)}, T_{c*}R\pi_*\mathcal{H}om^*(\mathcal{F}, \mathcal{F}))$ (see the argument above Remark 3.20), we know that

$$c(\mathcal{F}) \geq c'(\mathcal{F}).$$

Although $c(\mathcal{F})$ gives a potentially better estimation on capacity, it is sometimes easier to compute $c'(\mathcal{F})$ (see Example 4.7). Also, note that $c(\mathcal{F})$, if finite, provides a lower bound for the *boundary depth* (i.e., the length of the longest finite-length bar) of the sheaf barcode of $R\pi_*\mathcal{H}om^*(\mathcal{F}, \mathcal{F})$.

Example 4.5 Let $\mathcal{F} = \mathbf{k}_{[0,2)} \in \mathcal{T}(\text{pt})$. Then

$$R\pi_*\mathcal{H}om^*(\mathcal{F}, \mathcal{F}) = \mathcal{H}om^*(\mathcal{F}, \mathcal{F}) = \mathbf{k}_{[-2,0)}[1] \oplus \mathbf{k}_{[0,2)}$$

which is 2-torsion. Indeed, $c(\mathcal{F}) = c'(\mathcal{F}) = 2$.

Fig. 4.1 Deformation of a
torsion sheaf

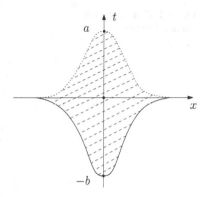

Example 4.6 Let $f : M \to \mathbb{R}$ be a differentiable function and $\mathcal{F} = \mathcal{F}_f$ be the canonical sheaf of f constructed in Example 3.9. By Example 3.21,

$$R\pi_* \mathcal{H}om^*(\mathcal{F}, \mathcal{F}) = \bigoplus \mathbf{k}_{(-\infty, 0)}.$$

Therefore, it is non-torsion and $c'(\mathcal{F}) = +\infty$.

Example 4.7 Let $\mathcal{F} := \mathbf{k}_Z$, where Z is depicted in Figure 4.1, as a deformation of Example 3.16 (smooth along $x = 0$). Then directly from the support of $\mathcal{F}$, we know that once $c \geq a + b$, $\mathcal{F} \to T_{c*}\mathcal{F}$ is the 0-map. Therefore,

$$c'(\mathcal{F}) \leq a + b.$$

In fact, $c'(\mathcal{F}) = a + b$. Indeed, if $c'(\mathcal{F}) < a + b$, then there exists some $c'(\mathcal{F}) < c < a + b$ such that $R\text{Hom}(\mathcal{F}, \mathcal{F}) \to R\text{Hom}(\mathcal{F}, T_{c*}\mathcal{F})$ is the zero morphism. In particular, $\mathbb{1}_{\mathcal{F}}$ maps to $\tau_c(\mathcal{F})$, which is the zero morphism. However, since $c < a + b$, there still exists a non-empty intersection between $\mathcal{F}$ and $T_{c*}\mathcal{F}$. By its definition, $\tau_c(\mathcal{F})$ is induced simply by a restriction, so $\tau_c(\mathcal{F}) \neq 0$. This gives a contradiction.

Due to the simplicity of $\mathcal{F}$ in this example, we can actually compute $c(\mathcal{F})$ explicitly. First of all, $R\pi_* \mathcal{H}om^*(\mathcal{F}, \mathcal{F}) = R\pi_*(\mathcal{F}^a *_{np} \mathcal{F})$ where $\mathcal{F}^a$ is the reflection of $\mathcal{F}$ with respect to x-axis, but still keeps the boundary open above and closed below. The main object we are interested in is then $\mathcal{F}^a *_{np} \mathcal{F}$. Moreover, for each $x \in \mathbb{R}_x$,

$$\mathcal{F}^a|_{\{x\} \times \mathbb{R}_t} = \mathbf{k}_{[-t_-^a(x), t_+^a(x))} \quad \text{and} \quad \mathcal{F}|_{\{x\} \times \mathbb{R}_t} = \mathbf{k}_{[-t_-(x), t_+(x))}.$$

Meanwhile, $t_-^a(x) = t_+(x)$ and $t_+^a(x) = t_-(x)$. Therefore, Figure 4.2 shows a stalk computational picture for the convolution $*_{np}$ over x. Therefore,

$$(\mathcal{F}^a *_{np} \mathcal{F})|_{\{x\} \times \mathbb{R}_t} = \mathbf{k}_{[0, t_-(x) + t_+(x))} \oplus \mathbf{k}_{[-t_-(x) - t_+(x), 0)}[-1].$$

Fig. 4.2 Computation of
maximal torsion

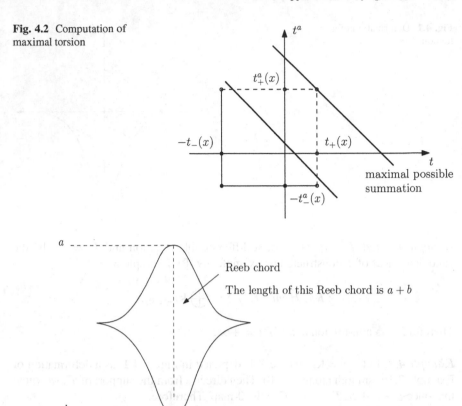

Fig. 4.3 Capacity characterized by a Reeb chord

The maximal $t_-(x)+t_+(x)$ we can get is $a+b$, from the fiber over $x = 0$. Therefore,

$$R\pi_*\mathcal{H}om^*(\mathcal{F},\mathcal{F}) = R\pi_*(\mathcal{F}^a *_{\mathrm{np}} \mathcal{F}) = \mathbf{k}_{[0,a+b)}[1] \oplus \mathbf{k}_{[-a-b,0)}.$$

Indeed, it is $(a + b)$-torsion.

Remark 4.7 The example above is very enlightening in the following sense. View $\partial \bar{Z}$ as the (x, t)-projection of a Legendrian knot K in $\mathbb{R}^3$ with the standard contact 1-form $\alpha = dt - ydx$. Note that y can be completely recovered from $\partial \bar{Z}$ via the relation $y = dt/dx$ (in particular, over $(0, a)$ and $(0, -b)$, $y = 0$). Moreover, with respect to α, the Reeb vector field is $\frac{\partial}{\partial t}$, so the Reeb chord of this K is shown in Figure 4.3. On the other hand, when we project K to the (x, y)-plane, we get an "∞"-figure picture as show in Figure 4.4. This can be viewed as an immersed Lagrangian manifold in $\mathbb{R}^2$ (cf. Example 3.16). One can check that the area of the shaded disk is also equal to $a + b$. In fact, if we put "+" to indicate the data from the upper half-curve and similarly for "−" to indicate the data from the lower half-curve, then

Fig. 4.4 Translation between Reeb chord and bounding disk

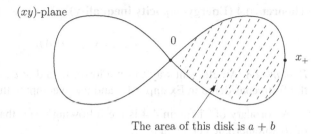

(xy)-plane

The area of this disk is $a + b$

$$\text{Area of disk} = \int_0^{x_+} \left(\frac{dt^-}{dx} - \frac{dt^+}{dx} \right) dx$$

$$= \int_0^{x_+} \frac{dt^-}{dx} dx - \int_0^{x_+} \frac{dt^+}{dx} dx$$

$$= (t(x_+)^- - t(0)^-) - (t(x_+)^+ - t(0)^+)$$

$$= (0 - (-b)) - (0 - a) = a + b.$$

Suggested by L. Polterovich, it would be interesting to generalize this observation to more complicated immersed Lagrangian manifolds, and define a capacity in terms of bounded disks. If this is difficult using the classical geometric methods, then sheaf method might serve as the right tool.

Definition 4.3 (Capacity of a domain) For a fixed domain $A \subset T^*M$, define

$$c_{\text{sheaf}}(A) = \sup\{c(\mathcal{F}) \mid \mathcal{F} \in \mathcal{T}_A(M)\}.$$

Example 4.8 Example 4.7 can be shrinked to any open ball, therefore, for any (small) open ball $B(\epsilon) \subset T^*M$, we know that $c_{\text{sheaf}}(B(\epsilon)) > 0$.

Recall that in Sect. 1.6 we have defined the displacement energy of a subset $A \subset T^*M$ by, $e(A) := \inf\{\|\phi\|_{\text{Hofer}} \mid \phi(A) \cap A = \emptyset\}$. Except for some trivial cases such as when A is non-displaceable (then $e(A) = +\infty$ by definition) and when $A = \{\text{pt}\}$ (then $e(A) = 0$, **Exercise**), it is in general difficult to compute the exact value of $e(A)$. Finding a meaningful lower bound of $e(A)$ becomes one of the central problems in symplectic geometry. Starting with the groundbreaking work of Hofer and Zehnder [29] and Lalonde and McDuff [35], various ways to obtain such lower bounds (see [19, 50, 56]) became available. All of them are derived using some hard machinery. The following theorem is one of the highlights of Tamarkin category theory, and provides a new way to get a lower bound of $e(A)$.

Theorem 4.4 (Energy-capacity inequality) *For any $A \subset T^*M$,*

$$c_{\text{sheaf}}(A) \leq 2e(A).$$

Remark 4.8 An efficient way to get a lower bound of $c_{\text{sheaf}}(A)$ is to use sheaves of the form considered in Example 4.7 and try to compute their capacities.

A corollary of Theorem 4.4 is the following result that was achieved by means of J-holomorphic curves in [42].

Corollary 4.3 *For any (small) open ball $B(\epsilon) \subset T^*M$, $e(B(\epsilon)) > 0$.*

The proof of Theorem 4.4 is a combination of various important results that have been elaborated so far.

Proof *(of Theorem 4.4)* We need to show that for any $\phi \in \text{Ham}(T^*M)$ such that $\phi(A) \cap A = \emptyset$ we have $c(\mathcal{F}) \leq 2\|\phi\|_{\text{Hofer}}$ for all $\mathcal{F} \in \mathcal{T}_A(M)$. By Lemma 4.3 and the Separation Theorem,

$$R\pi_* \mathcal{H}om^*(\mathcal{F}, \mathcal{K}(\phi)|_{s=1} \circ \mathcal{F}) = 0.$$

Then, by Remark 3.21 and Proposition 3.7,

$$
\begin{aligned}
c(\mathcal{F}) &= 2d_{\mathcal{T}(M)}(R\pi_*\mathcal{H}om^*(\mathcal{F}, \mathcal{F}), 0) \\
&= 2d_{\mathcal{T}(M)}(R\pi_*\mathcal{H}om^*(\mathcal{F}, \mathcal{F}), R\pi_*\mathcal{H}om^*(\mathcal{F}, \mathcal{K}(\phi)|_{s=1} \circ \mathcal{F})) \\
&\leq 2d_{\mathcal{T}(M)}(\mathcal{H}om^*(\mathcal{F}, \mathcal{F}), \mathcal{H}om^*(\mathcal{F}, \mathcal{K}(\phi)|_{s=1} \circ \mathcal{F})) \\
&\leq 2\left(d_{\mathcal{T}(M)}(\mathcal{F}, \mathcal{F}) + d_{\mathcal{T}(M)}(\mathcal{F}, \mathcal{K}(\phi)|_{s=1} \circ \mathcal{F})\right) \\
&\leq 2d_{\mathcal{T}(M)}(\mathcal{F}, \mathcal{K}(\phi)|_{s=1} \circ \mathcal{F}) \leq 2\|\phi\|_{\text{Hofer}}
\end{aligned}
$$

where the final step comes from the stability result Theorem 4.3.   □

Remark 4.9 To some extent, using torsion elements to define a capacity as above in Definition 4.3 is similar to using boundary depth to define a capacity (see Section 5.3 in [57]). In fact, Corollary 5.12 in [57] proves one version of energy-capacity inequality which looks comparable with the proof of Theorem 4.4.

4.4 Symplectic Ball-Projector (Following Chiu)

In this section, the (ideal) goal is to describe/obtain objects in $\mathcal{T}_U(M)$ when U is an open domain of T^*M. Since U is open, strictly speaking $\mathcal{T}_U(M)$ is not well-defined (cf. (2) in Definition 3.2). Formally, one defines

$$\mathcal{T}_U(M) := (\mathcal{T}_{T^*M \setminus U}(M))^{\perp}$$

where orthogonality "$\perp$" is taken in $\mathcal{T}(M)$. The concrete example we will work out in this section is that $M = \mathbb{R}^n$ and $U = B(r) := \{q^2 + p^2 < r^2\} \subset T^*\mathbb{R}^n$, where q is the position coordinate in $\mathbb{R}^n$ and p is the momentum coordinate.

Recall how we obtain objects in $\mathcal{T}(M)(= \mathcal{D}_{\{\tau \leq 0\}}(\mathbf{k}_{M \times \mathbb{R}})^{\perp, l})$. Starting from any object $\mathcal{F} \in \mathcal{D}(\mathbf{k}_{M \times \mathbb{R}})$, (3.3) tells us that convolution with the distinguished triangle

$$\mathbf{k}_{M \times [0,\infty)} \longrightarrow \mathbf{k}_{\{0\}} \longrightarrow \mathbf{k}_{M \times (0,\infty)}[-1]$$

splits an object $\mathcal{F}$ into two orthogonal parts, and $\mathcal{F} * \mathbf{k}_{M \times [0,\infty)} \in \mathcal{T}(M)$. We call this kind of distinguished triangle an *orthogonal splitting triangle*. Interestingly enough, we will obtain objects in $\mathcal{T}_U(M)$ also with the help of an orthogonal splitting triangle. The object in the title — symplectic ball projector — is a building block in this distinguished triangle, serving a role analogous to that of $\mathbf{k}_{M \times [0,\infty)}$ in the construction of $\mathcal{T}(M)$. Explicitly, one has the following important result (Theorem 3.11 in [10]).

Theorem 4.5 *There exists an orthogonal splitting triangle in* $\mathcal{D}(\mathbf{k}_{\mathbb{R}_1^n \times \mathbb{R}_2^n \times \mathbb{R}})$,

$$P_{B(r)} \longrightarrow \mathbf{k}_{\{(q_1, q_2, t) \mid q_1 = q_2; t \geq 0\}} \longrightarrow Q_{B(r)} \xrightarrow{+1} \qquad (4.15)$$

such that for any $\mathcal{F} \in \mathcal{T}(\mathbb{R}^n)$, $\mathcal{F} \bullet_{\mathbb{R}_1^n} P_{B(r)} \in \mathcal{T}_{B_r}(\mathbb{R}^n)$ *and* $\mathcal{F} \bullet_{\mathbb{R}_1^n} Q_{B(r)} \in \mathcal{T}_{T^*\mathbb{R}^n \setminus B_r}(\mathbb{R}^n)$.

Remark 4.10 Note that $\mathcal{T}_{B_r}(\mathbb{R}^n) \subset \mathcal{T}_{\bar{B}_r}(\mathbb{R}^n)$, so the difficult part in the proof of Theorem 4.5 is to establish the orthogonality.

Sometimes we need to modify $P_{B(r)}$ and $Q_{B(r)}$ to be "symmetric". Consider the map $\delta : \mathbb{R} \times \mathbb{R} \to \mathbb{R}$ given by $(t_1, t_2) \mapsto t_2 - t_1$. Denote $P(B(r)) := \delta^{-1} P_{B(r)}$. Applying δ^{-1} to (4.15), one gets

$$P(B(r)) \longrightarrow \mathbf{k}_{\{q_1 = q_2; t_2 \geq t_1\}} \to Q(B(r)) \xrightarrow{+1} \qquad (4.16)$$

an orthogonal splitting triangle in $\mathcal{D}(\mathbf{k}_{\mathbb{R}_1^n \times \mathbb{R}_1 \times \mathbb{R}_2^n \times \mathbb{R}_2})$, due to the following exercise.

Exercise 4.4 For any $\mathcal{F} \in \mathcal{T}(\mathbb{R}_1^n)$, $\mathcal{F} \bullet_{\mathbb{R}_1^n} P_{B(r)} = \mathcal{F} \circ_{\mathbb{R}_1^n \times \mathbb{R}_1} P(B(r))$. Hence $P(B(r))$ or $Q(B(r))$ can be regarded as a kernel.

Definition 4.4 $P_{B(r)}$ in (4.15) or $P(B(r))$ in (4.16) is called **the** symplectic ball-projector.

To justify this name, in Sect. 4.6 we will show that any such element in $\mathcal{D}(\mathbf{k}_{\mathbb{R}_1^n \times \mathbb{R}_2^n \times \mathbb{R}})$ which fits into an orthogonal splitting triangle such as (4.15) under "$\bullet$", or in $\mathcal{D}(\mathbf{k}_{\mathbb{R}_1^n \times \mathbb{R}_1 \times \mathbb{R}_2^n \times \mathbb{R}_2})$ which fits into an orthogonal splitting triangle such as (4.16) under "$\circ$", is unique up to isomorphism.

4.4.1   Construction of a Ball-Projector

We will give the explicit constructions of $P_{B(r)}$ and $Q_{B(r)}$ and leave to the interested reader to check the original proof of Theorem 4.5 in [10]. The constructions of $P_{B(r)}$ and $Q_{B(r)}$ are closely related to the Lagrangian submanifolds in Sect. 3.5. Recall how we obtain objects $\mathcal{F}$ in $\mathcal{T}_L(M)$ when L is a Lagrangian submanifold which admits a generating function. For instance (cf. Example 3.9), if

$$L = \text{graph}(df), \quad \text{then} \quad \mathcal{F}_f = \mathbf{k}_{\{(m,t) \mid f(m)+t \geq 0\}} \in \mathcal{T}_L(M). \tag{4.17}$$

Now, consider the function $H : T^*\mathbb{R}^n \to \mathbb{R}$, $(q, p) \mapsto q^2 + p^2$. Importance of this function H is explained by the following two facts.

(i) It defines $B(r)$ by $\{H < r^2\}$.
(ii) It generates an "simple" dynamics ϕ_H^a on $T^*\mathbb{R}^n$.

Here, $a \in \mathbb{R}$ stands for the time variable. For a fixed $a \in \mathbb{R}$, the Lagrangian submanifold $\text{graph}(\phi_H^a)$ in $T^*\mathbb{R}^n \times T^*\mathbb{R}^n$, up to a global Weinstein neighborhood identification, can be regarded as a Lagrangian submanifold of $T^*(T^*\mathbb{R}^n)$, denoted by Λ_a. By Proposition 9.33 in [36], there exists a generating function $S_a : T^*\mathbb{R}^n \to \mathbb{R}$ such that $\Lambda_a = \text{graph}(dS_a)$ for sufficiently small (non-zero) $a \in \mathbb{R}$. Let us temporarily ignore the extension issue on the time variable a (i.e., extending to the entire $\mathbb{R}$) in this section. Viewing a as an extra variable of the generating function, we can calculate this generating function $S : \mathbb{R} \times T^*\mathbb{R}^n \to \mathbb{R}$ explicitly, that is, for any $(q, p) \in T^*\mathbb{R}^n$ take as a starting point the formula

$$S(a, q, p) = \int_\gamma p\,dq - H\,da,$$

where γ is the Hamiltonian flow trajectory starting at (q, p) and flow for time a.

Exercise 4.5 For S defined above, check that

$$dS = -H\,da - p\,dq + (\phi_H^a)^*(p\,dq). \tag{4.18}$$

Note that $\text{graph}(dS)$ is a Lagrangian submanifold of $T^*\mathbb{R} \times T^*(T^*\mathbb{R}^n)$.

Remark 4.11 Here, we remark that viewing a as a dynamical variable is crucial, since its counterpart/co-vector H, the energy, gives a chance to control the domain algebraically. "Algebraically" means that one can use sheaf operators like composition and convolution to obtain the desired restriction of H (thus one gets a restriction on the choice of (q, p)).

Next, consider the transformation

$$\text{graph}(dS) \xrightarrow{\text{pre-quantize}} (\text{graph}(dS), -S) \xrightarrow{\text{conical lift}} \widehat{L}$$

where $\widehat{L} \subset T^*\mathbb{R} \times T^*(T^*\mathbb{R}^n) \times T^*_{\{\tau>0\}}\mathbb{R}$ is a homogenous Lagrangian submanifold. After a local change of variables (due to a twist condition satisfied by our H here, i.e., $\partial^2 H/\partial p^2 = 2 > 0$), $T^*\mathbb{R} \times T^*(T^*\mathbb{R}^n) \times T^*_{\{\tau>0\}}\mathbb{R} \simeq T^*_{\{\tau>0\}}(\mathbb{R} \times \mathbb{R}^n \times \mathbb{R}^n \times \mathbb{R})$, where the position coordinates are denoted by (a, q_1, q_2, t) and the momentum coordinates by (b, p_1, p_2, τ). From the left-hand side of (4.18) and (4.17) above, we know that there exists an element in $\mathcal{T}(\mathbb{R} \times \mathbb{R}^n \times \mathbb{R}^n)$,

$$\mathcal{F}_S := \mathbf{k}_{\{(a,q_1,q_2,t)\,|\,S_a(q_1,q_2)+t\geq 0\}}, \tag{4.19}$$

and it satisfies $SS(\mathcal{F}_S) = \widehat{L}$ modulo some 0-section part.

Remark 4.12 (1) One thing to be emphasized is that after changing variables from (q, p) to (q_1, q_2), the generating function S is no longer well-defined at some values of a (for instance $a = 0$). To overcome this difficulty, we need to extend $\mathcal{F}_S$ in a certain way so that $\mathcal{F}_S$ becomes well-defined over the entire $\mathbb{R}$ (see Page 619–620 in [10]). A more precise formula of $\mathcal{F}_S$ is provided in Sect. 4.5. Here, let us emphasize that

$$(\mathcal{F}_S)|_{a=0} := \mathbf{k}_{\{q_1=q_2;t\geq 0\}}.$$

(2) Note that, for the dynamics of the Hamiltonian flow ϕ_H^a, there are so far two ways to associate sheaves to it. One is by means of (4.19) and the other is the sheaf quantization from [26]. It would be interesting to compare these two methods in details (see Appendix A.3).

Definition 4.5 (H-ball-projector)

$$P_{B(r),H} = \mathcal{F}_S \bullet_a \mathbf{k}_{\{t+ab\geq 0\}}[1] \circ_b \mathbf{k}_{\{b<r^2\}} \in \mathcal{D}(\mathbf{k}_{\mathbb{R}_1^n \times \mathbb{R}_2^n \times \mathbb{R}})$$

and

$$Q_{B(r),H} = \mathcal{F}_S \bullet_a \mathbf{k}_{\{t+ab\geq 0\}}[1] \circ_b \mathbf{k}_{\{b\geq r^2\}} \in \mathcal{D}(\mathbf{k}_{\mathbb{R}_1^n \times \mathbb{R}_2^n \times \mathbb{R}})$$

where $\bullet_a$ means comp-convolution over $\mathbb{R}_a$ and $\circ_b$ means composition over $\mathbb{R}_b$.

Note that both $P_{B(r),H}$ and $Q_{B(r);H}$ are H-dependent, as their notations indicates. The logic to pass from $P_{B(r),H}$ to *the* ball-projector $P_{B(r)}$ is

$$P_{B(r),H} \xrightarrow{\;\;H\text{-independent}\;\;} P_{B(r)} \xrightarrow{\;\;\text{unique up to isom.}\;\;} P_{B(r)}$$

where both steps will be proved in Sect. 4.6.

Example 4.9 (Relation between $P_{B(r),H}$ and $Q_{B(r),H}$) Note that $P_{B(r),H}$ and $Q_{B(r),H}$ are related by the exact sequence $\mathbf{k}_{\{b<r^2\}} \to \mathbf{k}_{\mathbb{R}} \to \mathbf{k}_{\{b\geq r^2\}}$. In order to fit into the sequence $P_{B(r),H} \to \mathbf{k}_{\{q_1=q_2;t\geq 0\}} \to Q_{B(r),H} \xrightarrow{+1}$, we need to show that

$$\mathcal{F}_S \bullet_a \mathbf{k}_{\{t+ab\geq 0\}}[1] \circ_b \mathbf{k}_{\mathbb{R}} = \mathbf{k}_{\{q_1=q_2;t\geq 0\}}.$$

In fact, this comes from the relation $\mathbf{k}_{\{t+ab\geq 0\}} \circ_b \mathbf{k}_{\mathbb{R}} = \mathbf{k}_{\{a=0;t\geq 0\}}$ and then the fact that $(\mathcal{F}_S)|_{a=0} = \mathbf{k}_{\{q_1=q_2;t\geq 0\}})$. Here, we give the detailed computation of $\mathbf{k}_{\{t+ab\geq 0\}} \circ_b \mathbf{k}_{\mathbb{R}}$. By definition,

$$\mathbf{k}_{\{t+ab\geq 0\}} \circ_b \mathbf{k}_{\mathbb{R}} = Rp_! \mathbf{k}_{\{t+ab\geq 0\}}$$

where $p : \mathbb{R}_t \times \mathbb{R}_b \times \mathbb{R}_a \to \mathbb{R}_t \times \mathbb{R}_a$ is the projection. For any fixed $(t, a) \in \mathbb{R} \times \mathbb{R}$, we can solve for b. There are three cases:

- When $a < 0$, then $b \leq -t/a$, so compactly supported cohomology vanishes.
- When $a > 0$, then $b \geq -t/a$, so compactly supported cohomology vanishes.
- When $a = 0$, then $b \in \mathbb{R}$, so the compactly supported cohomology equals to $\mathbf{k}[-1]$.

Therefore, the assertion is proved.

Exercise 4.6 Show that

$$P_{B(r),H} = \mathcal{F}_S \bullet_a \mathbf{k}_{\{(a,t)\,|\,-t/r^2 \leq a \leq 0\}}.$$

In other words, check that $\mathbf{k}_{\{t+ab\geq 0\}}[1] \circ_b \mathbf{k}_{\{b<r^2\}} = \mathbf{k}_{\{(a,t)\,|\,-t/r^2 \leq a \leq 0\}}.$

We will end this section by clarifying the mysterious part $\bullet_a \mathbf{k}_{\{t+ab\geq 0\}}$ in Definition 4.5. This is called *Fourier-Sato transform* in our set-up. Let us use the following short subsection to get a better understanding of this operator.

4.4.2 Fourier-Sato Transform (Brief)

This operator can be illustrated by the following basic example. For more general result, see Lemma 3.7.10 in [32].

Example 4.10 Let $\mathcal{F} = \mathbf{k}_{[0,\infty)} \in \mathcal{D}(\mathbf{k}_{\mathbb{R}_a})$, where $\mathbb{R}_a$ denotes $\mathbb{R}$ in the coordinate a. Consider $\mathbf{k}_{\{ab\geq 0\}} \in \mathcal{D}(\mathbf{k}_{\mathbb{R}_a \times \mathbb{R}_b})$. We can compute

$$\mathcal{F} \circ_a \mathbf{k}_{\{ab\geq 0\}} = \mathbf{k}_{[0,\infty)} \circ_a \mathbf{k}_{\{ab\geq 0\}}$$
$$= Rp_!(\mathbf{k}_{[0,\infty)\times\mathbb{R}} \otimes \mathbf{k}_{\{ab\geq 0\}})$$
$$= \mathbf{k}_{(-\infty,0)}.$$

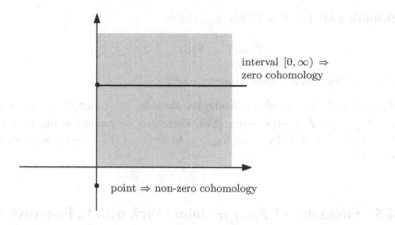

interval $[0, \infty) \Rightarrow$
zero cohomology

point $\Rightarrow$ non-zero cohomology

Fig. 4.5 Computation of the Sato-Fourier transform

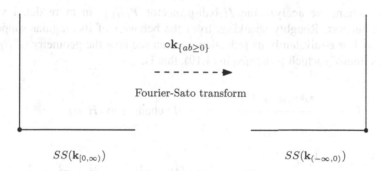

$\circ \mathbf{k}_{\{ab \geq 0\}}$

- - - - - - ➤

Fourier-Sato transform

$SS(\mathbf{k}_{[0,\infty)})$ $SS(\mathbf{k}_{(-\infty,0)})$

Fig. 4.6 Geometry of the Fourier-Sato transform

Figure 4.5 shows the computation where $H_c^*(\mathbb{R}, \mathbf{k}_{[0,\infty)}) = 0$ and $H_c^*(\mathbb{R}, \mathbf{k}_{\{pt\}}) = \mathbf{k}$. Denote the coordinates of $T^*\mathbb{R}_a$ by (a, A) and of $T^*\mathbb{R}_b$ by (b, B). Clearly,

$$SS(\mathcal{F}) = SS(\mathbf{k}_{[0,\infty)}) = \{(0, A) \mid A \geq 0\} \cup \{(a, 0) \mid a \geq 0\}$$

and

$$SS(\mathcal{F} \circ \mathbf{k}_{\{ab \geq 0\}}) = SS(\mathbf{k}_{(-\infty,0)}) = \{(b, 0) \mid b \leq 0\} \cup \{(0, B) \mid B \geq 0\}.$$

Define the anti-reflection $r : \mathbb{R} \times \mathbb{R} \to \mathbb{R} \times \mathbb{R}$ by $(x, y) \mapsto (-y, x)$, and observe that $SS(\mathcal{F} \circ \mathbf{k}_{\{ab \geq 0\}}) = r(SS(\mathcal{F}))$. Specifically, $(0, A) \mapsto (-A, 0)(\simeq (b, 0))$, where $-A, b \leq 0$ and $(a, 0) \mapsto (0, a)(\simeq (0, B))$, where $a, B \geq 0$ (see Figure 4.6).

Exercise 4.7 Let $\mathcal{F} = \mathbf{k}_{(0,\infty)}$. Then $\mathcal{F} \circ \mathbf{k}_{\{ab \geq 0\}} = \mathbf{k}_{[0,\infty)}[-1]$.

Definition 4.6 For $\mathcal{F} \in \mathcal{D}(\mathbf{k}_{M \times \mathbb{R}_a})$, define

$$\widehat{\mathcal{F}} = \mathcal{F} \circ_a \mathbf{k}_{\{ab \geq 0\}} \in \mathcal{D}(\mathbf{k}_{M \times \mathbb{R}_b});$$

$\widehat{\mathcal{F}}$ is called the *Fourier-Sato transform of* $\mathcal{F}$.

Remark 4.13 We can also consider the sheaf $\mathbf{k}_{\{ab \leq 0\}}$; then it is easy to check that $\widehat{\mathcal{F}} \circ \mathbf{k}_{\{ab \leq 0\}} = \mathcal{F}$ up to a degree shift. Therefore, we can define the *inverse Fourier-Sato transform* of $\mathcal{F}$ by using $\mathbf{k}_{\{ab \leq 0\}}$. In terms of the singular support, $(b, a) \to (a, -b)$.

4.5   Geometry of $P_{B(r),H}$ (Joint Work with L. Polterovich)

In this section, we analyze the H-ball-projector $P_{B(r),H}$ in more detail via its singular support. Roughly speaking, from the behavior of its singular support at level $\tau = 1$ or equivalently its reduction, we can see how the geometry of $P_{B(r),H}$ changes from $\mathcal{F}_S$ which is defined in (4.19), that is,

$$\widehat{L} \xrightarrow{\quad \bullet_a \mathbf{k}_{\{t + ab \geq 0\}}[1] \quad} (a, -H) \text{ changes to } (H, a)$$

$$\xrightarrow{\quad \circ_b \mathbf{k}_{\{b < r^2\}} \quad} \text{ restriction to } \{H < r^2\} \iff q^2 + p^2 < r^2.$$

Note that $H < r^2$ is exactly the desired restriction on (q, p) to obtain the ball $B(r)$. In this section, we will make the transform above more precise by doing detailed computations on singular supports. Moreover, we will give geometric interpretations of our computations. The material in this section, in particular Subsect. 4.5.2, is not completely included in [10].

4.5.1   Singular Support of $P_{B(r),H}$

First of all, due to the change of variables issue as mentioned in Remark 4.12, we need a more precise formula of $\mathcal{F}_S$ so that it becomes well-defined for all $a \in \mathbb{R}$. Given any time non-zero $a \in \mathbb{R}$, choose $M \gg 1$ such that a/M is sufficiently small (hence $S(a/M, \cdot, \cdot)$ is well-defined). Explicitly,

$$S(a/M, q_1, q_2) = \frac{q_1^2 + q_2^2}{2 \tan(2a/M)} - \frac{q_1 q_2}{\sin(2a/M)}. \tag{4.20}$$

Fig. 4.7 Closed cone from a ball-projector

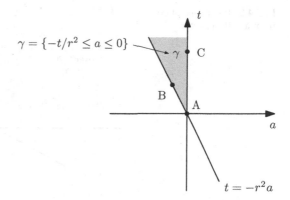

$$\gamma = \{-t/r^2 \le a \le 0\}$$

$t = -r^2 a$

Then one *defines*

$$(\mathcal{F}_S)|_a := \mathbf{k}_{\{S(a/M,q_1,q_2)+t\ge 0\}} \bullet_{\mathbb{R}_2^n} \mathbf{k}_{\{S(a/M,q_2,q_3)+t\ge 0\}} \bullet_{\mathbb{R}_3^n} \cdots \bullet_{\mathbb{R}_M^n} \mathbf{k}_{\{S(a/M,q_M,q_{M+1})+t\ge 0\}}$$

$$= R\rho_! \mathbf{k}_{\{(q_1,\dots,q_{M+1},t) \mid \sum S(a/M,q_i,q_{i+1})+t\ge 0\}}$$

where $\rho : (\mathbb{R}^n)^{M+1} \times \mathbb{R} \to \mathbb{R}^n \times \mathbb{R}^n \times \mathbb{R}$, $(q_1,\dots,q_{M+1},t) \mapsto (q_1,q_{M+1},t)$.

Exercise 4.8 Check that $\mathcal{F}_S$ is well-defined, that is, $(\mathcal{F}_S)|_a$ does not depend on M.

Repeatedly using the geometric meaning the of comp-convolution operator $\bullet$ from Exercise 3.2 (or directly applying the pushforward formula in Proposition 2.6), one sees that

$$SS(\mathcal{F}_S) \subset \left\{ \left((a, -\tau H), (q_1, -\tau p_1), \tau \phi_H^a(q_1, p_1), -\sum S(a/M, q_i, q_{i+1}), \tau \right) \mid \tau \ge 0 \right\}, \tag{4.21}$$

where $a \in \mathbb{R}$ and $(q_1,\dots,q_{M+1}) \in (\mathbb{R}^n)^{M+1}$. Second, by Exercise 4.6, we know that

$$SS(P_{B(r),H}) = SS(\mathcal{F}_S \bullet_a \mathbf{k}_{\{(a,t) \mid -t/r^2 \le a \le 0\}}).$$

Note that the set $\gamma := \{(a,t) \mid -t/r^2 \le a \le 0\}$ is a closed cone, as shown in Figure 4.7, Similarly to the computations in Sect. 3.11, we know that for $SS(\mathbf{k}_\gamma)$ there are three possible cases (ignoring the trivial case inside $\mathrm{int}(\gamma)$).

A. For $(0,0)$, the co-vector part of $SS(\mathbf{k}_\gamma)$ is polar cone γ° as in Figure 4.8.
B. For $(a, -r^2 a)$ with $a < 0$, the co-vector part of $SS(\mathbf{k}_\gamma)$ is $(r^2\tau, \tau)$ with $\tau \ge 0$.
C. For $(0,t)$ with $t > 0$, the co-vector part of $SS(\mathbf{k}_\gamma)$ is $(-\tau, 0)$ with $\tau \ge 0$.

Using Exercise 3.2, it is easy to check the following result.

Fig. 4.8 Polar cone of cone from a ball-projector

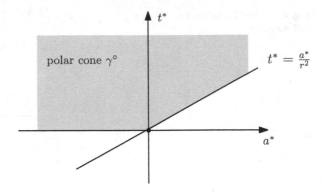

Proposition 4.5 *For the H-ball-projector $P_{B(r),H}$ defined in Definition 4.5, up to the 0-section part from the support of $P_{B(r),H}$,*

$$SS(P_{B(r),H}) \subset \begin{array}{l} \{(q_1, -\tau p_1, q_1, \tau p_1, 0, \tau) \mid q_1^2 + p_1^2 \leq r^2\} \cup \\ \{(q_1, -\tau p_1, \tau \phi_H^a(q_1, p_1), -S - r^2 a, \tau) \mid q_1^2 + p_1^2 = r^2, a < 0\}. \end{array}$$
(4.22)

Label the first component as part (i) *and the second component as part* (ii).

Exercise 4.9 Compute $SS(Q_{B(r),H})$.

Remark 4.14 Part (i) in (4.22) comes from the case A of $SS(\mathbf{k}_\gamma)$ and part (ii) from the case B of $SS(\mathbf{k}_\gamma)$. Case C of $SS(\mathbf{k}_\gamma)$ also provides some information, but by a dimension counting it is not coisotropic. Hence, by Theorem 2.8, it is not counted in $SS(P_{B(r),H})$.

The following example is very important (and readers are encouraged to check by themselves).

Example 4.11 Let $\mathcal{F} = \mathbf{k}_{\{f + t \geq 0\}}$ where $f : \mathbb{R}^n \to \mathbb{R}$ is a differentiable function. Then, up to the 0-section part from the support of $\mathcal{F} \bullet_{\mathbb{R}_1^n} P_{B(r),H}$,

$$SS(\mathcal{F} \bullet_{\mathbb{R}_1^n} P_{B(r),H}) \subset \begin{array}{l} \{(q_1, -\tau p_1, q_1, \tau p_1, 0, \tau) \mid q_1^2 + p_1^2 \leq r^2; p_1 = df(q_1)\} \cup \\ \{(q_1, -\tau p_1, \tau \phi_H^a(q_1, p_1), -S - r^2 a - f, \tau) \mid \\ \qquad q_1^2 + p_1^2 = r^2, p_1 = df(q_1), a < 0\}. \end{array}$$

Let us label the components in the right-hand side above in order as part (i) and part (ii). In other words, the operator $\bullet_{\mathbb{R}_1^n} P_{B(r),H}$ acts on $SS(\mathcal{F})$ in two steps: (i) cut and take only the part of $SS(\mathcal{F})$ inside $B(r)$ and (ii) flow along ϕ_H^a (for $a < 0$) for those parts of $SS(\mathcal{F})$ intersecting $\partial B(r)$ nontrivially.

4.5.2 Geometric Interaction with Projectors

We will give two geometric interpretations of the computational result in Example 4.11. In fact, the only mysterious aspect is part (ii) in Example 4.11 which happens on the boundary $\partial B(r)$.

The first explanation is based on the theory of generating functions. Recall that $F : M \times \mathbb{R}^K \to \mathbb{R}$ generates the manifold

$$L = \left\{ \left(m, \frac{\partial F}{\partial m}(m, \xi) \right) \, \middle| \, \frac{\partial F}{\partial \xi}(m, \xi) = 0 \right\}$$

by viewing $\xi \in \mathbb{R}^K$ as a "ghost variable".

Exercise 4.10 Generically, L is an immersed Lagrangian submanifold in T^*M.

Then the lift of L defined by $\widehat{L} = \{(\tau L, -F, \tau) \mid \tau > 0\} \subset T^*_{\{\tau > 0\}}(M \times \mathbb{R})$ is a homogeneous Lagrangian submanifold. For $\mathcal{F} = \mathbf{k}_{\{f+t \geq 0\}}$, where $f : \mathbb{R}^n_1 \to \mathbb{R}$ is a differentiable function, consider the function $F \colon \mathbb{R}_a \times \mathbb{R}^n_1 \times \mathbb{R}^n_2 \to \mathbb{R}$ given by $F(a, q_1, q_2) := S(a, q_1, q_2) + r^2 a + f(q_1)$. View $\mathbb{R}_a \times \mathbb{R}^n_1$ as "ghost variables". Recall that the derivatives of S are,

$$\frac{\partial S}{\partial a} = -H, \quad \frac{\partial S}{\partial q_1} = -p_1 \quad \text{and} \quad \frac{\partial S}{\partial q_2} = p_2.$$

Now take $\xi = (a, q_1)$, then $\{\frac{\partial F}{\partial \xi} = 0\}$ splits into the following two cases,

$$\frac{\partial F}{\partial a} = 0 \Longleftrightarrow -H + r^2 = 0 \Longrightarrow \text{restriction of } (q_1, p_1) \text{ to } \partial B(r)$$

and

$$\frac{\partial F}{\partial q_1} = 0 \Longleftrightarrow -p_1 + df(q_1) = 0 \Longrightarrow \text{intersection with } SS(\mathcal{F}).$$

Therefore, part (ii) in Example 4.11 is just the conical lift $\widehat{L}$ given by the generating function $F = S + r^2 a + f$.

The second explanation reveals a deep geometry behind the expression of part (ii) in Example 4.11. To start, we recall some general geometric terms. Let $\Sigma \subset (\mathbb{R}^{2n}, \omega_{\text{std}})$ be a hypersurface. The distribution $L_z = \{J_0 v \mid v \perp T_z \Sigma, z \in \Sigma\}$ gives the (1-dim) characteristic foliation of Σ. In particular, if $\Sigma = H^{-1}(r)$ is a regular level set of a function H, then $X_H(z) \in L_z$ (so the integral curves of $X_H(z)$ correspond to the characteristic foliation). More importantly,

Exercise 4.11 Different choices of such H for which $\Sigma = H^{-1}(r)$ only result in reparametrizations of integral curves. In other words, the characteristic foliation of Σ is independent of the function that defines it (if exists). Then computing the characteristic foliation reduces to computing the Hamiltonian flow of *any* preferred defining Hamiltonian.

Moreover, observe that a hypersurface of $\mathbb{R}^{2n}$ can be lifted to a conical hypersurface in $T^*_{\{\tau>0\}}(\mathbb{R}^n \times \mathbb{R})$. Therefore, it is important to understand how the Hamiltonian dynamics behaves in this homogenized space. Here are some general computations. Let $H(q, p)$ be a Hamiltonian function on T^*M. Lift H to $T^*_{\{\tau>0\}}(M \times \mathbb{R})$ by the rule

$$\widehat{H}(q, \xi, t, \tau) = \tau H\left(q, \frac{\xi}{\tau}\right), \qquad (4.23)$$

where (q, ξ, t, τ) are the coordinates of $T^*(M \times \mathbb{R})$, and $p = \xi/\tau$. Take the symplectic form $\omega = dq \wedge d\xi + dt \wedge d\tau$ on $T^*M \times T^*\mathbb{R}_{>0}$. Then the Hamiltonian equations of $\widehat{H}$ with respect ω on $T^*_{\{\tau>0\}}(M \times \mathbb{R})$ read,

$$\begin{cases} \dot{q} = \dfrac{\partial H}{\partial q}, \\[2mm] \dot{\xi} = -\tau \dfrac{\partial H}{\partial q}, \\[2mm] \dot{t} = H - \dfrac{\xi}{\tau}\dfrac{\partial H}{\partial p}, \\[2mm] \dot{\tau} = 0. \end{cases} \qquad (4.24)$$

Example 4.12 For $H(q, p) = q^2 + p^2$ on $\mathbb{R}^2$, $\widehat{H} = \tau(q^2 + (\xi/\tau)^2)$. On the boundary of ball at level $\tau = 1$,

$$t(a) \sim O(a).$$

See Figure 4.9 for the Hamiltonian dynamics in $T^*_{\{\tau>0\}}(\mathbb{R}^2 \times \mathbb{R})$. Once we fix a starting level τ, the dynamics remains on this level τ (because $\dot{\tau} = 0$). But, more interestingly, the direction t keeps tracking the evolution in terms of the negative symplectic action functional!

For $\partial B(r) \subset T^*\mathbb{R}^n$, its lift is a homogeneous hypersurface in $T^*_{\{\tau>0\}}(\mathbb{R}^n \times \mathbb{R})$, namely

$$\widehat{\Sigma} = \{\tau(q^2 + (\xi/\tau)^2) = r^2\} = \widehat{H}^{-1}(r^2).$$

Moreover, by (4.24) and Exercise 4.11, the Hamiltonian vector field gives the characteristic foliation. Therefore, back to Example 4.11,

$$\text{part (ii)} \cap \{\tau = 1\} = \bigcup_{x \in \mathrm{graph}(df) \cap \partial B(r)} \begin{array}{l} \text{(truncated) leaf of characteristic} \\ \text{foliation of } \partial B(r) \text{ containing } x. \end{array} \qquad (4.25)$$

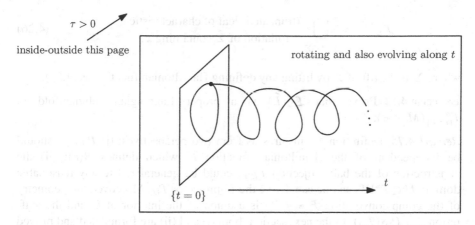

Fig. 4.9   Hamiltonian dynamics lifted to the homogenized space

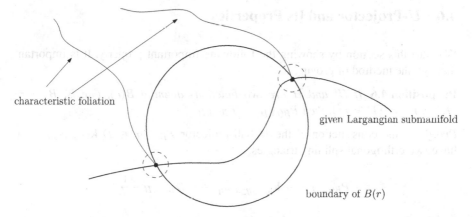

Fig. 4.10 Union of some leaves of the characteristic foliation

This is illustrated in Figure 4.10. Part (ii) is just the union of the sets (4.25) for all $\tau \in \mathbb{R}_{>0}$ (with rescaled graph(df) and $\partial B(r)$). The advantage of the expression for (4.25) is that it describes part (ii) without using any Hamiltonian function (mainly thanks to Exercise 4.11). Moreover, this can be generalized to the following geometric operation associated to any reasonable hypersurface $\Sigma \subset \mathbb{R}^{2n}$.

Definition 4.7 Fix a hypersurface $\Sigma \subset \mathbb{R}^{2n}$ such that Σ can be realized as a regular level set. Define the *leaf-union operator* $\mathcal{L}_\Sigma$ for any $\widehat{L}$ a conical Lagrangian submanifold in $T^*_{\{\tau>0\}}(\mathbb{R}^n \times \mathbb{R})$ by,

$$\mathcal{L}_{\Sigma}(\widehat{L}) = \bigcup_{x \in \widehat{L} \cap \widehat{\Sigma}} \begin{array}{l} \text{(truncated) leaf of characteristic} \\ \text{foliation of } \widehat{\Sigma} \text{ containing } x, \end{array} \qquad (4.26)$$

where $\widehat{\Sigma}$ is the lift of Σ by lifting any defining Hamiltonian function as (4.23).

Exercise 4.12 Prove that $\mathcal{L}_{\Sigma}(\widehat{L})$ is a proper Lagrangian submanifold in $T^*_{\{\tau > 0\}}(M \times \mathbb{R})$.

Remark 4.15 Definition 4.7 inspires us from two perspectives: (i) $P_{B(r),H}$ should be independent of the Hamiltonian function H which defines $B(r)$; (ii) the construction of the ball-projection $P_{B(r)}$ could be generalized to any reasonable domain $U \subset T^*\mathbb{R}^n$, so one can define the U-projector P_U. Moreover, the geometry of the comp-convolution $\mathcal{F} \bullet_{\mathbb{R}^n_1} P_U$ is a union of the interior of U and the leaf-union $\mathcal{L}_{\partial U}(SS(\mathcal{F}))$. In the next section, both (i) and (ii) are formulated and proved accurately.

4.6 U-Projector and Its Properties

We start this section by showing the following important property. It is important through the method of proving it.

Proposition 4.6 *If H' and H are two functions defining $B(r)$, that is, $B(r) = \{H' < r^2\} = \{H < r^2\}$, then $P_{B(r),H'} \simeq P_{B(r),H}$.*

Proof By the construction of the H-ball-projector $P_{B(r),H} \in \mathcal{D}(\mathbf{k}_{\mathbb{R}^n_1 \times \mathbb{R}^n_2 \times \mathbb{R}})$, we have two orthogonal splitting triangles,

$$P_{B(r),H} \longrightarrow \mathbf{k}_{\{q_1 = q_2; t \geq 0\}} \longrightarrow Q_{B(r),H} \xrightarrow{+1}$$

and

$$P_{B(r),H'} \longrightarrow \mathbf{k}_{\{q_1 = q_2; t \geq 0\}} \longrightarrow Q_{B(r),H'} \xrightarrow{+1}$$

Applying $P_{B(r),H'} \bullet_{\mathbb{R}^n_2}$ to the first distinguished triangle, one gets

$$P_{B(r),H'} \bullet_{\mathbb{R}^n_2} P_{B(r),H} \longrightarrow P_{B(r),H'} \longrightarrow P_{B(r),H'} \bullet_{\mathbb{R}^n_2} Q_{B(r),H} \xrightarrow{+1}.$$

Then, in view of Corollary 3.1, we know that for any $\mathcal{F} \in \mathcal{T}(\mathbb{R}^n)$,

$$\mathcal{F} \bullet_{\mathbb{R}^n_1} P_{B(r),H'} \bullet_{\mathbb{R}^n_2} Q_{B(r),H} = (\mathcal{F} \bullet_{\mathbb{R}^n_1} P_{B(r),H'}) \bullet_{\mathbb{R}^n_2} Q_{B(r),H} = 0,$$

which implies that $P_{B(r),H'} \bullet_{\mathbb{R}^n_2} Q_{B(r),H} = 0$. Then $P_{B(r),H'} \bullet_{\mathbb{R}^n_2} P_{B(r),H} \simeq P_{B(r),H'}$. Similarly, applying $\bullet_{\mathbb{R}^n_2} P_{B(r),H}$ to the second distinguished triangle, one gets that $P_{B(r),H'} \bullet_{\mathbb{R}^n_2} P_{B(r),H} \simeq P_{B(r),H}$. This completes the proof. $\qquad \square$

Note that the only property we used in the argument above is the *orthogonality* from the orthogonal splitting triangle that defines $P_{B(r),H}$, which makes this argument quite formal. The same argument also establishes the uniqueness of the ball-projectors (see the following exercise).

Exercise 4.13 If both $P_{B(r)}$ and $P'_{B(r)}$ are ball-projectors, then $P_{B(r)} \simeq P'_{B(r)}$.

Next we move to general domains. We refer to Definition 4.1 in [10].

Definition 4.8 We call an open domain $U \subset \mathbb{R}^{2n}$ *admissible* if there exists an orthogonal splitting triangle in $\mathcal{D}(\mathbf{k}_{\mathbb{R}_1^n \times \mathbb{R}_2^n \times \mathbb{R}})$

$$P_U \longrightarrow \mathbf{k}_{\{q_1 = q_2; t \geq 0\}} \longrightarrow Q_U \xrightarrow{+1},$$

that is, for any $\mathcal{F} \in \mathcal{T}(\mathbb{R}_1^n)$, $\mathcal{F} \bullet_{\mathbb{R}_1^n} P_U \in \mathcal{T}_U(\mathbb{R}^n)$ and $\mathcal{F} \bullet_{\mathbb{R}_1^n} Q_U \in \mathcal{T}_{T^*\mathbb{R}^n \setminus U}(\mathbb{R}^n)$. This P_U or $P(U) := \delta^{-1} P_U$ is called *the U-projector*.

Example 4.13 (Examples of admissible domains) Consider an open domain U such that $U = \{H < r\}$ as a regular sublevel set (and then $\partial \bar{U} = \{H = r\}$) for some H on $\mathbb{R}^{2n}$. For instance, take the ball $B(r)$ or the ellipsoid $E(r_1, \ldots, r_n)$. Indeed, we can carry out the same procedure as in the previous sections. Construct $P_{U,H}$ first by choosing a preferred defining function of U, and then prove that $P_{U,H}$ is independent of H (so can be denoted by P_U) by the orthogonality property as in Proposition 4.6. Finally, as in Exercise 4.13, P_U is unique up to isomorphism.

The following lemma lists some crucial functorial properties of the U-projector.

Lemma 4.4 *Let U, V be admissible domains of $\mathbb{R}^{2n}$.*

(1) *For the inclusion $i : V \hookrightarrow U$, there exists a well-defined map $i_* : P(V) \to P(U)$. Moreover, this association is functorial.*
(2) *For any Hamiltonian isotopy $\phi = \{\phi_s\}_{s \in I}$ on $\mathbb{R}^{2n}$,*

$$P(\phi_s(U)) := \mathcal{K}^{-1}(\phi)|_s \circ P(U) \circ \mathcal{K}(\phi)|_s$$

is the $\phi_s(U)$-projector, where $\mathcal{K}(\phi)$ is the Guillermou-Kashiwara-Schapira's sheaf quantization associated to the homogeneous lift of the Hamiltonian isotopy ϕ. In other words, U is admissible if and only if $\phi_s(U)$ is admissible.

Proof (1) We have a distinguished triangle

$$P(U) \longrightarrow \mathbf{k}_{\{q_1 = q_2; t_2 \geq t_1\}} \longrightarrow Q(U) \xrightarrow{+1}.$$

Apply $\circ P(V)$, where "$\circ$" is a short notation for $\circ_{\mathbb{R}^n \times \mathbb{R}}$, and one gets

$$P(U) \circ P(V) \longrightarrow P(V) \longrightarrow Q(U) \circ P(V) \xrightarrow{+1}.$$

Then $Q(U) \circ P(V) = 0$, which implies $P(U) \circ P(V) \simeq P(V)$. On the other hand, one has a morphism $P(V) \to \mathbf{k}_{\{q_1 = q_2; t_2 \geq t_1\}}$. Applying $P(U) \circ$, we get $P(U) \circ P(V) \to P(U)$. Therefore,

$$P(V) \simeq P(U) \circ P(V) \to P(U).$$

(2) We only need to show that, for any $\mathcal{G} \in \mathcal{T}_{\mathbb{R}^n \setminus \phi_s(U)}(\mathbb{R}^n)$, $\mathrm{RHom}(\mathcal{F} \circ P(\phi_s(U)), \mathcal{G}) = 0$. Indeed,

$$\mathrm{RHom}(\mathcal{F} \circ P(\phi_s(U)), \mathcal{G}) = \mathrm{RHom}(\mathcal{F} \circ \mathcal{K}^{-1}(\phi)|_s \circ P(U) \circ \mathcal{K}(\phi)|_s, \mathcal{G})$$

$$= \mathrm{RHom}(\mathcal{F} \circ \mathcal{K}^{-1}(\phi)|_s \circ P(U), \mathcal{G} \circ \mathcal{K}^{-1}(\phi)|_s) = 0,$$

since $\mathcal{G} \circ \mathcal{K}^{-1}(\phi)|_s \in \mathcal{T}_{\mathbb{R}^n \setminus U}(\mathbb{R}^n)$ while $\mathcal{F} \circ \mathcal{K}^{-1}(\phi)|_s \circ P(U) = (\mathcal{F} \circ \mathcal{K}^{-1}(\phi)|_s) \circ P(U) \in \mathcal{T}_U(\mathbb{R}^n) = \mathcal{T}_{\mathbb{R}^n \setminus U}(\mathbb{R}^n)^{\perp}$.

$\square$

Exercise 4.14 Complete the proof above by verifying the functorial property of i_*, i.e., for any inclusion $U \xrightarrow{i} V \xrightarrow{j} W$, we have $j_* \circ i_* = (j \circ i)_*$.

Remark 4.16 (1) The proof of (1) in Lemma 4.4 also shows when $V = U$, $P(U) \circ P(U) \simeq P(U)$. This justifies the name "projector". (2) In general, for two admissible domains V and U with inclusion $i : V \hookrightarrow U$, it seems difficult to describe explicitly the induced map $i_* : P(V) \to P(U)$. However, for some special cases, for instance, when U is an admissible domain as in Example 4.13, and $V = cU$ with $0 < c \leq 1$, a rescaling of U, one knows precisely what i_* is. In fact, suppose that $U = \{H < 1\}$; since P_{cU} is independent of the choice of a defining function, we can still take H, and then $cU = \{H < c\}$. By the construction of P_U (similarly to Definition 4.5),

$$P_U \simeq \mathbf{k}_{\{S+t \geq 0\}} \bullet_a \mathbf{k}_{\{(a,t) \,|\, -t \leq a \leq 0\}} \quad \text{and} \quad P_{cU} \simeq \mathbf{k}_{\{S+t \geq 0\}} \bullet_a \mathbf{k}_{\{(a,t) \,|\, -t/c \leq a \leq 0\}}.$$

Note that the first part $\mathbf{k}_{\{S+t \geq 0\}}$ is the same for both P_U and P_{cU}, due to the same choice of H (and the generating function S depends only on H). The difference only comes from the cone part. For the inclusion $i : cU \hookrightarrow U$, the map $i_* : P_{cU} \to P_U$ is induced by a morphism

$$\mathbf{k}_{\{(a,t) \,|\, -t/c \leq a \leq 0\}} \longrightarrow \mathbf{k}_{\{(a,t) \,|\, -t \leq a \leq 0\}}, \tag{4.27}$$

and this morphism is induced by the morphism ι in the following commutative diagram guaranteed by the triangulated structure:

$$
\begin{array}{ccccc}
\mathbf{k}_{\{b < c\}} & \longrightarrow & \mathbf{k}_{\mathbb{R}} & \longrightarrow & \mathbf{k}_{\{b \geq c\}} \xrightarrow{+1} \\
\downarrow \iota & & \downarrow 1 & & \downarrow rmres \\
\mathbf{k}_{\{b < 1\}} & \longrightarrow & \mathbf{k}_{\mathbb{R}} & \longrightarrow & \mathbf{k}_{\{b \geq 1\}}. \xrightarrow{+1}
\end{array}
$$

Exercise 4.15 Prove $\mathrm{Cone}(\iota) = \mathbf{k}_{[c,1)}[-1]$.

4.7   Sheaf Barcode from a U-Projector

To start this section, let us consider the new object

$$S_T(U) := R\mathrm{Hom}(P(U), \mathbf{k}_{\{q_1=q_2;t_2-t_1\geq T\}}) \tag{4.28}$$

for any fixed constant $T \geq 0$. For $S_T(U)$, here is a remarkable observation from Definition 4.6 in [10] that reduces considerably the computational difficulty via the following successive deductions:

$$
\begin{aligned}
S_T(U) :&= R\mathrm{Hom}(\delta^{-1}P_U, \mathbf{k}_{\{q_1=q_2;t_2-t_1\geq T\}}) \\
&= R\mathrm{Hom}(P_U, R\delta_*\mathbf{k}_{\{q_1=q_2;t_2-t_1\geq T\}}) \\
&= R\mathrm{Hom}(P_U, \mathbf{k}_{\{q_1=q_2;t\geq T\}}) && t,\ \text{the coordinate of } \mathbb{R} \\
&= R\mathrm{Hom}(P_U, R\Delta_*\mathbf{k}_{\mathbb{R}^n \times \{t\geq T\}}) && \Delta : \mathbb{R}^n \to \mathbb{R}^n \times \mathbb{R}^n, \ \text{diagonal emb.} \\
&= R\mathrm{Hom}(\Delta^{-1}P_U, \mathbf{k}_{\mathbb{R}^n \times \{t\geq T\}}) \\
&= R\mathrm{Hom}(\Delta^{-1}P_U, \pi^{-1}\mathbf{k}_{\{T\geq T\}}) && \pi : \mathbb{R}^n \times \mathbb{R} \to \mathbb{R}, \ \text{projection} \\
&= R\mathrm{Hom}(\Delta^{-1}P_U, \pi^!\mathbf{k}_{\{t\geq T\}}[-n]) \\
&= R\mathrm{Hom}(R\pi_!\Delta^{-1}P_U, \mathbf{k}_{\{t\geq T\}}[-n]).
\end{aligned}
$$

Upshot: Denote $\mathcal{F}(U) := R\pi_!\Delta^{-1}P_U$; thus, we have successfully transferred the discussion of $P(U)$ to $\mathcal{F}(U)$, which is a constructible sheaf over $\mathbb{R}$. For $\mathcal{F}(U)$, we can read its information from its sheaf barcode.

Example 4.14 Since the singular support of a constructible sheaf over $\mathbb{R}$ has a particularly simple structure, we can calculate $SS(\mathcal{F}(U))$ to see when there are non-trivial fibers, and this will be the only interesting part of a constructible sheaf over $\mathbb{R}$. Let us focus on $U = B(r)$. By (4.22), we already have a meaningful upper bound of $SS(P_U)$. The following diagram defines $\mathcal{F}(U)$:

$$
\begin{array}{ccc}
\mathbb{R}^n \times \mathbb{R}^n \times \mathbb{R} & \xleftarrow{\ \Delta\ } & \mathbb{R}^n \times \mathbb{R} \\
& & \downarrow{\scriptstyle \pi} \\
& & \mathbb{R}.
\end{array}
$$

By functorial properties of the singular support, specifically, Propositions 2.6 and 2.7, one can compute $SS(\mathcal{F}(U)) = SS(R\pi_! \Delta^{-1} P_U)$. Let us do this step by step. We have

$$SS(\Delta^{-1} P_U) = \left\{ (q, p, t, \tau) \in T^*_{\{\tau > 0\}}(\mathbb{R}^n \times \mathbb{R}) \left| \begin{array}{l} \exists (q_1, -\tau p_1, q_2, \tau p_2, t, \tau) \in SS(P_U) \\ \quad \text{and} \quad \Delta(q, t) = (q_1, q_2, t) \\ \quad \text{and} \quad \Delta^*(-\tau p_1, \tau p_2, \tau) = (p, \tau) \end{array} \right. \right\}.$$

Note that

$$\Delta(q, t) = (q, q, t) = (q_1, q_2, t) \quad \Longrightarrow \quad q_1 = q_2 (= q)$$

which also implies $p_1 = p_2 (= p')$. Therefore,

$$\Delta^*(-\tau p_1, \tau p_2, \tau) = \Delta^*(-\tau p', \tau p', \tau) = (-\tau p' + \tau p', \tau) = (0, \tau) \quad \Longrightarrow \quad p = 0.$$

Now, $SS(P_U)$ is contained the union of two parts (see (4.22)). Due to the restriction from part (i), we should take $(q, 0, 0, \tau)$ for any $(q, 0)$ such that $q^2 \leq r^2$. Due to the restriction from part (ii), there are multiple times a such that $(q_2, p_2) = \phi^a_H(q_1, p_1) = (q_1, p_1)$, that is $a = n\pi$ for $n \in \mathbb{Z}_{<0}$. In other words, this corresponds to the fixed points of the flow ϕ^a_H. Then we should take

$$(q, 0, -S(n\pi, q, p) - r^2(n\pi), \tau) \quad \text{s.t.} \quad (q, p) \text{ is fixed and } q^2 + p^2 = r^2.$$

Now let us compute $-S(n\pi, q, p) - r^2(n\pi)$. Recall the (global) formula for S, that is $S(a, p, q) = \int_\gamma p \, dq - H \, da$, where γ is the Hamiltonian trajectory starting at (q, p) flowing for time a. Since H is constant along the orbits of the Hamiltonian flow,

$$-S(a, q, p) - r^2 a = -\int_\gamma p \, dq.$$

Parametrizing $(p, q) = (r \cos(\theta_0 + 2a), r \sin(\theta_0 + 2a))$, we have

$$-S(n\pi, q, p) - r^2(n\pi) = -2r^2 \int_0^{n\pi} \cos^2(\theta_0 + 2a) \, da$$

$$= -2r^2 \int_0^{n\pi} \frac{1 + \cos(2\theta_0 + 4a)}{2} \, da = -n\pi r^2.$$

Therefore,

$$SS(\Delta^{-1} P_U) \subset \{(q, 0, n\pi r^2, \tau) \mid q^2 \leq r^2, \tau \geq 0, n \in \mathbb{Z}_{\geq 0}\}.$$

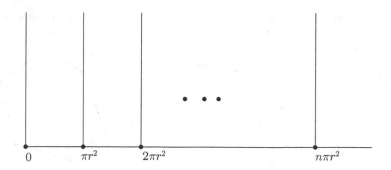

Fig. 4.11 Singular support of $\mathcal{F}(B(r))$

Furthermore,

$$SS(R\pi_!\Delta^{-1}P_U) \subset \left\{ (t, \tau) \in T^*\mathbb{R}_{\geq 0} \;\middle|\; \begin{array}{c} \exists (q, p, t, \tau) \in SS(\Delta^{-1}P_U) \\ \text{and } \pi(q, t) = t \\ \text{and } \pi^*(\tau) = (p, \tau) \end{array} \right\}.$$

Note that $\pi^*(\tau) = (0, \tau)$ and $t = n\pi r^2$ for $n \in \mathbb{Z}_{\geq 0}$. All in all, we get

$$SS(\mathcal{F}(U)) \subset \{(t, \tau) \in T^*\mathbb{R}_{\geq 0} \mid t = n\pi r^2\} \cup 0_{\mathbb{R}_{\geq 0}}. \tag{4.29}$$

Figure 4.11 shows the upper bound in (4.29).

The computation in Example 4.14 is not sufficient for describing $\mathcal{F}(U)$ exactly, since SS cannot tell the degrees and multiplicities. We need additional work to obtain an exact formula for $\mathcal{F}(U)$. However, the computation above reveals that, for a general admissible domain U, the non-trivial fibers of $SS(\mathcal{F}(U))$ occur at the symplectic actions of closed Hamiltonian loops.

Here is a more accurate description of $\mathcal{F}(U)$ when $U = B(r)$.

Lemma 4.5 *For any $T \geq 0$, the stalk of $\mathcal{F}(B(r))$ is*

$$\mathcal{F}(B(r))_T = H_c^*(Y_{T/(r^2M)}; \mathbf{k}),$$

where

$$Y_{T/(r^2M)} = \left\{ (q_1, \ldots, q_M) \in (\mathbb{R}^n)^M \;\middle|\; \sum_{i=1}^M S\left(\frac{-T}{r^2M}, q_i, q_{i+1}\right) \geq 0 \right\},$$

and M is sufficiently large such that $S(-T/(r^2M), \cdot, \cdot)$ is well-defined. Here we regard $(q_1, \ldots, q_M)$ as a discrete closed loop in the space $(\mathbb{R}^n)^{M+1}$, viewing it as $(q_1, \ldots, q_{M+1})$ where $q_1 = q_{M+1}$.

discrete Hamiltonian trajectory   discrete Hamiltonian loop

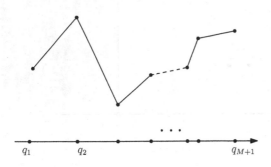
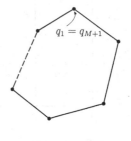

Fig. 4.12 Discrete Hamiltonian trajectory and loop

Remark 4.17 Since $\sum S(-T/(r^2 M), q_i, q_{i+1})$ is a differentiable function on the topological space $(\mathbb{R}^n)^M$, a finite-dimensional approximation of the loop space of $\mathbb{R}^n$, $Y_{T/(r^2 M)}$ is just a sublevel set and then $H_c^*(Y_{T/(r^2 M)}, \mathbf{k})$ is generated by the critical points of $\sum S(-T/(r^2 M), q_i, q_{i+1})$. Recall that a critical point of this function is a *discrete* Hamiltonian trajectory. Since we always identify q_1 (start point) with q_{M+1} (end point), generators of $H_c^*(Y_T; \mathbf{k})$ are *discrete* Hamiltonian loops of period T/r^2, see Figure 4.12.

Example 4.15 Let $U = B(r)$ and take $H = q^2 + p^2$. If T is sufficiently small, we can take $M = 1$. Recall the local expression of S in (4.20). Let $a = -T/r^2$. Then the condition above in Y_{T/r^2} transfers to

$$\left(\frac{1}{\tan(2a)} - \frac{1}{\sin(2a)} \right) q^2 \geq 0.$$

Since our a is negative and sufficiently close to 0,

$$\frac{1}{\tan(2a)} - \frac{1}{\sin(2a)} = \frac{\cos(2a) - 1}{\sin(2a)} \geq 0.$$

There is no constraint on q. Therefore, $\mathcal{F}(B(r))_T = H_c^*(\mathbb{R}^n, \mathbf{k}) = \mathbf{k}[-n]$.

With the same T as above, we can also choose $M = 2$, and then the condition above in $Y_{T/(2r^2)}$ says that

$$\frac{1}{\tan(a)}(q_1^2 + q_2^2) - \frac{2q_1 q_2}{\sin(a)} \geq 0,$$

where $a = -T/(2r^2)$. Note that we can normalize this quadratic form so that $Q_1 = q_1 - q_2$ and $Q_2 = q_1 + q_2$, that is, $\lambda_1(a)Q_1^2 + \lambda_2(a)Q_2^2 \geq 0$, where the eigenvalues λ_1 and λ_2 are

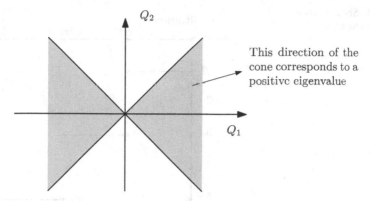

Fig. 4.13   Restriction imposed by the generating function

$$\lambda_1(a) = \frac{1}{\tan(a)} - \frac{1}{\sin(a)} \quad \text{and} \quad \lambda_2(a) = \frac{1}{\tan(a)} + \frac{1}{\sin(a)}.$$

Then $\lambda_1 \geq 0$ and $\lambda_2 \leq 0$. The constraint on Q_1 and Q_2 is represented by Figure 4.13. Note that this is *proper* homotopic to $\mathbb{R}^n$ (representing Q_1), so $\mathcal{F}(B(r))_T = H_c^*(Y_{T/(2r^2)}, \mathbf{k}) = H_c^*(\mathbb{R}^n; \mathbf{k}) = \mathbf{k}[-n]$.

This works for any choice of M and the answer is still $\mathbf{k}[-n]$. In general, given any $T \geq 0$, choosing any sufficiently large M and letting $a = -T/(r^2 M)$, the quadratic form constraint is normalized to be

$$\lambda_1(a) Q_1^2 + \cdots \lambda_M(a) Q_M^2 \geq 0,$$

where

$$\lambda_i(a) = \frac{1}{\tan(2a/M)} - \frac{\cos(2\pi(i-1)/M)}{\sin(2a/M)}.$$

Then counting the maximal possible number of positive eigenvalues, we get that $\# = 2\lceil \frac{T}{\pi r^2} \rceil - 1$. Note that this is independent of M! Each corresponding direction provides n free dimensions. Hence, $\mathcal{F}(B(r))_T = \mathbf{k}[n - 2n\lceil \frac{T}{\pi r^2} \rceil]$. Thus we obtain the sheaf barcode of $\mathcal{F}(B(r))$ shown in Figure 4.14.

Exercise 4.16 Prove that $\mathcal{F}(E(r, R, \ldots, R))_T = \mathbf{k}[n - 2(n-1)\lceil \frac{T}{\pi R^2} \rceil - 2\lceil \frac{T}{\pi r^2} \rceil]$.

We end this section with a technical part:

***Proof** (of Lemma 4.5)* For any $T \in \mathbb{R}$,

$$\mathcal{F}(B(r))_T = (R\pi_! \Delta^{-1} P_{B(r)})_T = H_c^*(\mathbb{R}^n, (\Delta^{-1} P_{B(r)})|_{\mathbb{R}^n \times \{T\}}).$$

Fig. 4.14 Sheaf barcode associated to $B(r)$

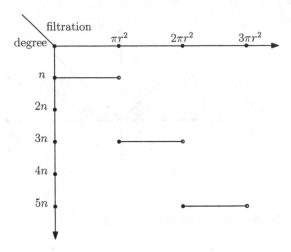

Recall the construction of $P_{B(r)} \in \mathcal{D}(\mathbf{k}_{\mathbb{R}_1^n \times \mathbb{R}_2^n \times \mathbb{R}})$:

$$P_{B(r)} = R\rho_! \mathbf{k}_{\{(a,q_1,\dots,q_{M+1},t) \mid \sum S(a/M,q_i,q_{i+1})+t \geq 0\}} \bullet_a \mathbf{k}_{\{(a,t) \mid -t/r^2 \leq a \leq 0\}}.$$

Since the diagonal embedding Δ does not involve the t-variable, we can restrict on $t = T$ first for $P_{B(r)}$, that is,

$$(P_{B(r)})|_T = R\rho_! \mathbf{k}_{\{(a,q_1,\dots,q_{M+1}) \mid \sum S(a/M,q_i,q_{i+1}) \geq 0\}} \circ_a \mathbf{k}_{\{a \mid -T/r^2 \leq a \leq 0\}}.$$

$$= R\rho_! \mathbf{k}_{\{(q_1,\dots,q_{M+1}) \mid \sum S(a/M,q_i,q_{i+1}) \geq 0 \text{ for } a \in [-T/r^2, 0]\}} := R\rho_! \mathbf{k}_{X_T}.$$

Now, consider the commutative diagram

$$
\begin{array}{ccc}
(\mathbb{R}^n)^{M+1} & \xrightarrow{\rho} & \mathbb{R}^n \times \mathbb{R}^n \\
\bar{\Delta} \uparrow & & \uparrow \Delta \\
(\mathbb{R}^n)^M & \xrightarrow{\bar{\rho}} & \mathbb{R}^n
\end{array}
$$

where $\bar{\rho}(q_1,\dots,q_M) = q_1$ and $\bar{\Delta}(q_1,\dots,q_M) = (q_1,\dots,q_M,q_{M+1})$. The base change formula says that

$$\Delta^{-1} R\rho_! \mathbf{k}_{X_T} = R\bar{\rho}_! \bar{\Delta}^{-1} \mathbf{k}_{X_T} := R\bar{\rho}_! \mathbf{k}_{\bar{Y}_T}.$$

Therefore,

$$(\Delta^{-1} P_{B(r)})|_{\mathbb{R}^n \times \{T\}} = \left\{ q_1 \in \mathbb{R}^n \;\middle|\; \begin{array}{l} \exists \text{ a loop } (q_1,\dots,q_{M+1}=q_1) \text{ s.t.} \\ \sum S(a,q_i,q_{i+1}) \geq 0 \text{ for } a \in [-T/(r^2 M), 0] \end{array} \right\}.$$

For a computational perspective,

$$\mathcal{F}(B(r))_T = H^*_c(\mathbb{R}^n, R\bar{\rho}_! \mathbf{k}_{\bar{Y}_T}) = H^*_c((\mathbb{R}^n)^M, \mathbf{k}_{\bar{Y}_T}) = H^*_c(\bar{Y}_T; \mathbf{k}),$$

where

$$\bar{Y}_T = \left\{ (q_1, \ldots, q_M) \in (\mathbb{R}^n)^M \;\middle|\; \sum S(a, q_i, q_{i+1}) \geq 0 \text{ for } a \in [-T/(r^2 M), 0] \right\}.$$

Exercise 4.17 Check that when $a \in [-T/(r^2 M), 0]$, the sets

$$Y_{-a} = \left\{ (q_1, \ldots, q_M) \in (\mathbb{R}^n)^M \;\middle|\; \sum S(a, q_i, q_{i+1}) \geq 0 \right\}$$

are nested, i.e., for any $a \leq a'$, $Y_{-a'} \subset Y_{-a}$. Then $\bar{Y}_T = \bigcup_{a \in [-T/(r^2 M), 0]} Y_{-a} = Y_{T/(r^2 M)}$. Note that Example 4.15 supports this conclusion.

This completes the proof.   □

4.8 Comparison with Symplectic Homology

Symplectic homology is a powerful tool for studying domains in a symplectic manifold. It has a rich history and various versions (see [4, 17, 60]). Although the definition of the U-projector is quite abstract and given in the language of sheaves, in this section we exhibit some common features shared by $\mathcal{F}(U)$ and the symplectic homology of U, conventionally denoted by SH(U). We will briefly recall the construction of symplectic homology, but only focus on the case of ball $B(r)$.

Symplectic homology of $B(r)$. Consider the function H shown in Figure 4.15, radial symmetric on $\mathbb{R}^{2n}$. The standard Hamiltonian Floer theory (see [2, 16, 47]) computes the filtered Floer homology HF(H), which roughly speaking, counts *1-periodic orbits* of the Hamiltonian flow of H, up to homology. Here is an important exercise.

Exercise 4.18 For a radially symmetric H in Figure 4.15 (represented by the thick solid line segments, smoothed near $\partial B(r)$), denote the slope of the part inside $B(r)$ by k_H. Prove that the 1-periodic orbits of the Hamiltonian flow of H appear near $\partial B(r)$, and they are in a one-to-one correspondence with the slopes $n\pi$ within $[0, k_H]$. Moreover, the associated action (for positive time) is the y-intercept of the line with the slope $n\pi$. More precisely,

$$\text{the action of the orbit corresponding to the slope } n\pi = -n\pi r^2.$$

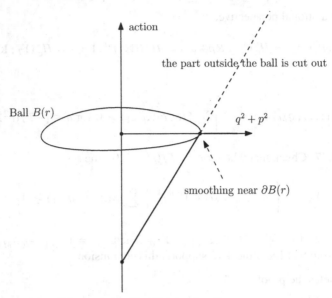

A radial symmetric Hamiltonian function

Fig. 4.15 Radial symmetric Hamiltonian associated to $B(r)$

In order to associate an object to $B(r)$, we need to eliminate the dependence of Hamiltonian functions. The procedure of defining symplectic homology consists of the following three steps: (1) consider a family of functions H as in Figure 4.15 with increasing slopes; (2) compute the Hamiltonian Floer homology for each one of them; (3) take a limit to get an algebraic object associated only to $B(r)$. This algebraic object is called *the symplectic homology of* $B(r)$, and is denoted by

$$SH(B(r)) := \varinjlim_{i \to \infty} HF(H_i).$$

For a general U, this procedure applies and one gets $SH(U)$. It is an interesting fact that $SH(U)$ is independent of the sequence of Hamiltonians, as long as it eventually blows up or dominating. Usually, a specific Hamiltonian function is fixed for any practical computation, say $H(q, p) = q^2 + p^2 - r^2$. Then take the sequence $\{\lambda H\}_{\lambda \in [1,\infty)}$ (see Figure 4.16).

Observations and comparisons. By a simple computation,

$$\phi_{\lambda H}^a = \begin{pmatrix} \cos(2\lambda a) & \sin(2\lambda a) \\ -\sin(2\lambda a) & \cos(2\lambda a) \end{pmatrix} = \phi_H^{\lambda a}.$$

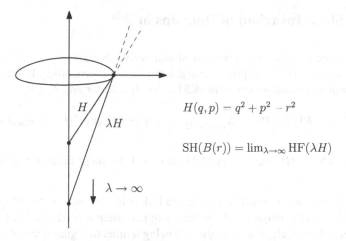

$$H(q,p) = q^2 + p^2 - r^2$$

$$SH(B(r)) = \lim_{\lambda \to \infty} HF(\lambda H)$$

Fig. 4.16 Dominating Hamiltonians for computing $SH(B(r))$

Observe that a 1-periodic orbit of the flow $\phi^a_{\lambda H}$ is an $a\lambda$-periodic orbit of the flow ϕ^a_H. Therefore, rescaling a Hamiltonian function is (dynamically) equivalent to rescaling time. Up to homology, $SH(B(r))$ does in fact count *all* periodic orbits of a *fixed* Hamiltonian function. Then, by the conclusion from Exercise 4.18, rescaling H results in more choices of $n\pi$. Then, up to a factor,

$$\text{the restriction of time} = \text{the restriction of action.} \qquad (4.30)$$

Here is a list of similarities between $\mathcal{F}(B(r))$ and $SH(B(r))$ that follow directly from their constructions.

- Both constructions *cut* Hamiltonian functions outside $B(r)$.
- Both constructions are *independent* of the Hamiltonian functions defining them.
- Both constructions can be *generalized* to other reasonable domains of $\mathbb{R}^{2n}$.
- Up to homology, both constructions count the *Hamiltonian closed trajectories*. For $\mathcal{F}(B(r))$, one counts *discrete* closed trajectories via discrete approximations of the loop space; for $SH(B(r))$, one counts *smooth* closed trajectories via Floer theory.
- Both constructions admit *filtered versions*, where filtrations are determined by symplectic actions T. Although for $\mathcal{F}(B(r))$ the filtration is determined by time via the computation in Lemma 4.5, relation (4.30) shows that this is equivalent the filtration by symplectic actions.

4.9   A Sheaf Invariant of Domains in $\mathbb{R}^{2n}$

In this section, we define an invariant of admissible domains $U \subset \mathbb{R}^{2n}$ via the U-projector P_U or $\mathcal{F}(U)$, and prove several key functorial properties. In fact, we have already seen this construction from (4.31). Recall that, for any $T \geq 0$,

$$S_T(U) := R\mathrm{Hom}(P(U), \mathbf{k}_{\{q_1=q_2; t_2-t_1 \geq T\}}) = R\mathrm{Hom}(\mathcal{F}(U), \mathbf{k}_{[T,\infty)}[-n]). \tag{4.31}$$

Definition 4.9 $S_T(U)$ defined in (4.31) is called the *sheaf invariant of U at level* T.

View $\mathcal{F}(U)$ as a sheaf barcode, and denote the set of jump points by $\mathrm{Spec}\,(\mathcal{F}(U))$. In the language of the theory of persistent $\mathbf{k}$-modules, $\{S_T(U)\}_{T\geq 0}$ is a persistence $\mathbf{k}$-module, thanks to the following lemma that guarantees the structure maps.

Lemma 4.6 *For any $T_1 \leq T_2$, there exists a natural map $\iota_{T_1,T_2} : S_{T_1}(U) \to S_{T_2}(U)$. Moreover, if $[T_1, T_2] \cap \mathrm{Spec}\,(\mathcal{F}(U)) = \emptyset$, then ι_{T_1,T_2} is an isomorphism.*

Proof For $T_1 \leq T_2$, the restriction $\mathbf{k}_{[T_1,\infty)} \xrightarrow{\mathrm{res}} \mathbf{k}_{[T_2,\infty)}$ induces ι_{T_1,T_2}. Moreover, the mapping cone of ι_{T_1,T_2} is

$$R\mathrm{Hom}(\mathcal{F}(U), \mathbf{k}_{[T_1,T_2)}[-n]).$$

Assume that $\mathcal{F}(U) = \bigoplus \mathbf{k}_{[a,b)}$. Then our hypothesis implies that

$$R\mathrm{Hom}(\mathcal{F}(U), \mathbf{k}_{[T_1,T_2)}[-n]) = \bigoplus_{[T_1,T_2)\subset[a,b)} R\mathrm{Hom}(\mathbf{k}_{[a,b)}, \mathbf{k}_{[T_1,T_2)}[-n])$$

for those $[a, b)$ in the summand of $\mathcal{F}(U)$ such that $[T_1, T_2] \subset [a, b)$. This is zero by Corollary A.2.   □

Example 4.16 By Example 4.15 and the computation of $R\mathrm{Hom}$, one gets $S_T(B(r)) = \mathbf{k}[-2mn]$ for $T \in [m\pi r^2, (m + 1)\pi r^2)$. For any $T_1 \leq T_2 \in (m\pi r^2, (m+1)\pi r^2)$, ι_{T_1,T_2} is an isomorphism, since $\mathrm{Spec}\,(\mathcal{F}(B(r))) = \{n\pi r^2 \mid n \in \mathbb{Z}_{\geq 0}\}$.

The following proposition is essentially a consequence of Lemma 4.4.

Proposition 4.7 *Let U, V be admissible domains of $\mathbb{R}^{2n}$. The following basic functorial properties hold.*

(1) *For any inclusion $i : V \hookrightarrow U$, there exists a well-defined functorial map $i^* : S_T(U) \to S_T(V)$. Moreover, i^* commutes with ι_{T_1,T_2}.*

(2) *For any Hamiltonian diffeomorphism ϕ on $T^*\mathbb{R}^n$, $S_T(\phi(U)) \simeq S_T(U)$.*

Proof (1) This comes from (1) in Lemma 4.4. We only need to check that i^* commutes with res_{T_1,T_2}. In fact, i^* is induced by $\circ i_*$ and restriction ι_{T_1,T_2} is induced by $\mathrm{res}_{T_1,T_2}\circ$, where they act on different sides. (2) This comes from (2) in Lemma 4.4. In fact,

$$S_T(\phi(U)) = R\mathrm{Hom}(P(\phi(U)), \mathbf{k}_{\{q_1=q_2;t_2-t_1\geq T\}})$$

$$= R\mathrm{Hom}(\mathcal{K}^{-1}(\phi)|_{s=1} \circ P(U) \circ \mathcal{K}(\phi)|_{s=1}, \mathbf{k}_{\{q_1=q_2;t_2-t_1\geq T\}})$$

$$= R\mathrm{Hom}(P(U), \mathcal{K}(\phi)|_{s=1} \circ \mathbf{k}_{\{q_1=q_2;t_2-t_1\geq T\}} \circ \mathcal{K}^{-1}(\phi)|_{s=1})$$

$$= R\mathrm{Hom}(P(U), \mathbf{k}_{\{q_1=q_2;t_2-t_1\geq T\}}) = S_T(U),$$

where the third equality comes from the fact that $\mathcal{K}(\phi)|_{s=1}\circ$ (or $\circ\mathcal{K}^{-1}(\phi)|_{s=1}$) is an automorphism of $\mathcal{T}(\mathbb{R}^n)$. $\qquad\square$

Remark 4.18 Note that symplectic homology $SH(U)$ can also be viewed as a persistence $\mathbf{k}$-module and satisfies these functorial properties from Proposition 4.7.

Example 4.17 Based on Remark 4.16, we have an explicit description of the induced map $i^* : S_T(B(r)) \to S_T(cB(r))$ for $0 < c \leq 1$. Note that this i^* is eventually induced by the restriction $\mathbf{k}_{\{b\geq cr^2\}} \to \mathbf{k}_{\{b\geq r^2\}}$ through the steps

$$\mathbf{k}_{\{b<cr^2\}} \longrightarrow \mathbf{k}_{\{b<r^2\}} \Longrightarrow P_{cU} \longrightarrow P_U \Longrightarrow \mathcal{F}(cU) \longrightarrow \mathcal{F}(U) \Longrightarrow S_T(U) \longrightarrow S_T(cU).$$

One way to understand i^* is to consider its mapping cone, which is eventually induced by the mapping cone of the restriction $\mathbf{k}_{\{b\geq cr^2\}} \to \mathbf{k}_{\{b\geq r^2\}}$, that is, $\mathbf{k}_{[cr^2,r^2)}$. The following exercise is a direct modification of Lemma 4.5.

Exercise 4.19 For any $T \geq 0$, fix M such that both $T/(cr^2 M)$ and $T/(r^2 M)$ are sufficiently small. Then

$$\mathrm{Cone}(\mathcal{F}(cU) \to \mathcal{F}(U))_T = H_c^*(Y_{T/(cr^2 M)}, Y_{T/(r^2 M)}; \mathbf{k}).$$

Since the right-hand side is a relative cohomology, it actually computes the compactly supported cohomology of an inter-level subset on $(\mathbb{R}^n)^M$ with respect to the function $S(a, \cdot, \cdot)$. In the case $U = B(r)$, we can determine this inter-level subset precisely thanks to the concrete formula (4.20) for the generating function.

Recall that the computation of $H_c^*(Y_{T/(r^2 M)}; \mathbf{k})$ reduces to counting the number of positive eigenvalues of a certain quadratic form. When the value T/r^2 increases, more positive eigenvalues appear. Denote

$$\{\text{positive eigenvalues up to } T/(cr^2)\} = \{\lambda_1, \ldots, \lambda_{m_1}, \ldots, \lambda_{m_c}\}$$

where $\lambda_1, \ldots, \lambda_{m_1}$ are those to T/r^2. From Figure 4.13, each λ_i where $i \in \{1, \ldots, m_1\}$ provides n free dimensions. Therefore, its corresponding inter-level subset is shown in Figure 4.17.

Fig. 4.17 Inter-level subset
computing mapping cone

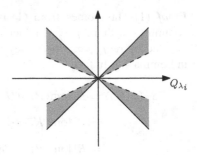

Exercise 4.20 Check that the direction corresponding to λ_i for $i \in \{1, \ldots, m_1\}$ in this inter-level subset provides zero compactly supported cohomology. Therefore,

$$\text{Cone}(\mathcal{F}(cB(r)) \to \mathcal{F}(B(r)))_T = \mathbf{k}[(m_1 - m_c)n], \qquad (4.32)$$

which is contributed by the newly appeared eigenvalues λ_j for $j \in \{m_1 + 1, \ldots, m_c\}$.

4.10   Proof of the Gromov Non-squeezing Theorem

In this section, we give a proof of the celebrated Gromov non-squeezing theorem based on the sheaf invariant developed in the previous sections. The proof given here is parallel to the proof via the theory of persistence $\mathbf{k}$-module (see Section 9.6 in [43]). Since a symplectic cylinder $Z(r)$ can be approximated by an ellipsoid $E(r, R, \ldots, R)$ with a sufficiently large R, we will prove the following (equivalent) statement.

Theorem 4.6 *If there exists a symplectic embedding* $\phi : B(r_1) \to E(r_2, R, \ldots, R)$ *(with* $R \gg \max\{r_1, r_2\}$*), then* $r_1 \leq r_2$.

Proof Suppose $r_1 > r_2$. Extend ϕ to a compactly supported Hamiltonian diffeomorphism Φ on $\mathbb{R}^{2n} (\simeq T^*\mathbb{R}^n)$. Choose $R_\dagger$ sufficiently large so that $B(R_\dagger)$ contains the support of Φ. Then the inclusion relation $\Phi(B(r_1)) \subset E(r_2, R, \ldots, R) \subset B(R_\dagger)$, yields has the following commutative diagram, for any $T \geq 0$:

$$
\begin{array}{ccccc}
S_T(\Phi(B(r_1))) & \xleftarrow{\ j_2^*\ } & S_T(E(r_2, R, ..., R)) & \xleftarrow{\ j_1^*\ } & S_T(B(R_\dagger)) \\
\ \ \downarrow{\scriptstyle\simeq} & & & & \ \ \downarrow{\scriptstyle j^*} \\
S_T(B(r_1)) & \xleftarrow{\hspace{3cm}=\hspace{3cm}} & & & S_T(B(r_1))
\end{array}
$$

where j^* is induced by the inclusion $j : B(r_1) \hookrightarrow B(R_+)$. Choose $T \in (\pi r_2^2, \pi r_1^2)$. Then

$$S_T(\Phi(B(r_1))) \simeq S_T(B(r_1)) = \mathbf{k}.$$

However, since $(R >)T > \pi r_2^2$, Exercise 4.16 shows that

$$\mathcal{F}(E(r_2, R, \ldots, R))_T = \mathbf{k}[n - 2(n - 1) - 4] = \mathbf{k}[-n - 2].$$

Therefore, $S_T(E(r_2, R, \ldots, R)) = \mathbf{k}[2]$, which implies that $j_2^* \circ j_1^* = 0$. On the other hand, since $S_T(B(R_+)) = \mathbf{k}$, the proof will be finished if we show $j^* \neq 0$. In fact, if $j^* = 0$, then $\mathrm{Cone}(j^*) \simeq \mathbf{k}[1] \oplus \mathbf{k}$ simply by the definition of mapping cone. In particular, it has rank 2. On the other hand, j^* is also induced by (4.32) which always has rank no greater than 1. Thus we get a contradiction! $\square$

Appendix A
Supplements

In this appendix, we provide details on the following three subjects. In Section A.1, the relation between persistence modules and constructible sheaves over $\mathbb{R}$. This explains why the category of persistence modules can be viewed as a special case of a Tamarkin category. In Section A.2, a detailed computation of the sheaf hom between two sheaves $\mathbf{k}_I$ and $\mathbf{k}_J$, where I, J are intervals of $\mathbb{R}$. This result helps us quickly read "sheaf barcodes", and has been used many times in the previous chapters. A.3 A dynamical interpretation of the Guillermou-Kashiwara-Schapira sheaf quantization from the perspective of semi-classical analysis. It turns out that the geometry (singular support) of the sheaf quantization can be regarded, to some extent, as a semi-classical counterpart of the classical Lagrangian suspension.

A.1 Persistence k-Modules vs. Sheaves

In this section, we exhibit a canonical correspondence between persistence $\mathbf{k}$-modules and constructible sheaves over $\mathbb{R}$. We say that V is in $(-, -]$-type if each bar in $\mathcal{B}(V)$ is of the form $(\cdot, \cdot]$. Similarly, we say a constructible sheaf $\mathcal{F}$ over $\mathbb{R}$ is in $(-, -]$-type if, by the decomposition theorem (Theorem 1.17 in [34]), each bar in $\mathcal{B}(\mathcal{F})$ is of the form $(\cdot, \cdot]$. It is in the same way to define objects in $[-, -)$-type. The main result is the following.

Theorem A.1 *Denote by $\mathcal{P}$ the category of persistence $\mathbf{k}$-modules in $(-, -]$-type and by $\mathrm{Sh}_c(\mathbb{R})$ the category of constructible sheaves over $\mathbb{R}$ in $(-, -]$-type. Then there exists an equivalence*

$$\Phi : \mathcal{P} \simeq \mathrm{Sh}_c(\mathbb{R}).$$

Moreover, for any $V \in \mathcal{P}$, $\mathcal{B}(V) = \mathcal{B}(\Phi(V))$ and for any $\mathcal{F} \in \mathrm{Sh}_c(\mathbb{R})$, $\mathcal{B}(\Psi(\mathcal{F})) = \mathcal{B}(\mathcal{F})$, where Ψ is the inverse of Φ.

© Springer Nature Switzerland AG 2020
J. Zhang, *Quantitative Tamarkin Theory*, CRM Short Courses,
https://doi.org/10.1007/978-3-030-37888-2

We call W an *anti-persistence* **k**-*module* if W has a persistence **k**-module structure but its transfer map goes in the wrong way, that is, $\iota_{t,s} : W_t \to W_s$ for any $s \le t$. By reparametrizing $\mathbb{R}$ by $t \to -t$, Theorem A.1 immediately gives the following result.

Corollary A.1 *Denote by* $\overline{\mathcal{P}}$ *the category of anti-persistence* **k**-*modules of* $[-, -)$-*type and by* $\overline{\mathrm{Sh}_c(\mathbb{R})}$ *the category of constructible sheaves over* $\mathbb{R}$ *of* $[-, -)$-*type. Then there exists an equivalence*

$$\overline{\Phi} : \overline{\mathcal{P}} \simeq \overline{\mathrm{Sh}_c(\mathbb{R})}.$$

Moreover, for any $W \in \overline{\mathcal{P}}$, $\mathcal{B}(W) = \mathcal{B}(\overline{\Phi}(W))$ *and for any* $\mathcal{G} \in \overline{\mathrm{Sh}_c(\mathbb{R})}$, $\mathcal{B}(\overline{\Psi}(\mathcal{G})) = \mathcal{B}(\mathcal{G})$, *where* $\overline{\Psi}$ *is the inverse of* $\overline{\Phi}$.

Note that the example computed in Sect. 3.9 involving $\mathcal{F}_f$ and $\mathcal{F}_g$ is an example of Corollary A.1, where the $[-, -)$-type is given by the defining property of Tamarkin category.

In the rest of this section, we will prove Theorem A.1 by constructing $\Phi : \mathcal{P} \to \mathrm{Sh}_c(\mathbb{R})$, and leave other routine checking as an exercise. The conclusion on the identification of barcodes comes from exactly the same argument as in Sect. 3.9 due to Lemma 3.5. As an agenda, we will construct such Φ by the following steps, where each notation that appears will be defined explicitly later:

$$\text{per. mod} \xrightarrow{\clubsuit} \text{sheaf on } \mathbb{R}_A \xrightarrow{\heartsuit} \text{sheaf on } \mathbb{R}_\gamma \xrightarrow{\spadesuit} \text{sheaf on } \mathbb{R}.$$

♣ In an abstract language, a persistence **k**-module can be viewed (or defined) as a functor $V : (\mathbb{R}, \le) \to \mathrm{Vect}$, where $(\mathbb{R}, \le)$ is $\mathbb{R}$ with the order $\le$ and Vect is the category of finite-dimensional vector spaces. As studied in [13], Sections 4.2.1 and 4.2.2, we can transfer such a functor to a sheaf over $\mathbb{R}$, but with a special topology defined as follows.

Definition A.1 Denote by $\mathbb{R}_A$ the real line $\mathbb{R}$ with the topology in which the open subsets are

$$\mathrm{Op}(\mathbb{R}_A) = \{[a, \infty), (b, \infty) \mid a, b \in \mathbb{R}\}.$$

Note that a basis of this topology is the collection $\{[a, \infty) \mid a \in \mathbb{R}\}$, since any $(b, \infty) = \bigcup_n [b + 1/n, \infty)$. This topology is called the *Alexandrov topology*.

For a given functor $V : (\mathbb{R}, \le) \to \mathrm{Vect}$, define a pre-sheaf $\mathcal{F}$ by,

$$\mathcal{F}([a, \infty)) = V_a \quad \text{and} \quad \mathcal{F}((b, \infty)) = \varprojlim_{n \to \infty} \mathcal{F}([b + 1/n, \infty)).$$

Exercise A.1 Check that $\mathcal{F}$ is a pre-sheaf, i.e., the restriction maps are well defined. Moreover, check that $\mathcal{F}$ is actually a sheaf (see Theorem 4.2.10 in [13]).

♡ Recall that the γ-topology on $\mathbb{R}$ with $\gamma = [0, \infty)$ consists of the open subsets of the form (a, ∞) (see Section 1.2 in [34]). Denote $\mathbb{R}_\gamma$ as $\mathbb{R}$ with the γ-topology. Then

$$\phi(= 1) : \mathbb{R}_A \longrightarrow \mathbb{R}_\gamma$$

is a continuous map and set $\mathcal{G} := \phi_* \mathcal{F}$. For each open subset (a, ∞) in $\mathbb{R}_\gamma$,

$$\mathcal{G}((a, \infty)) = (\phi_* \mathcal{F})((a, \infty)) = \mathcal{F}(\phi^{-1}(a, \infty)) = \mathcal{F}((a, \infty)),$$

and for the stalk at a,

$$\mathcal{G}_a = \varinjlim_{\epsilon \to 0} \mathcal{G}((a - \epsilon, \infty)) = \varinjlim_{\epsilon \to 0} \mathcal{F}((a - \epsilon, \infty)).$$

Exercise A.2 Check that $\mathcal{G}_a = V_a$ for any $a \in \mathbb{R}$ (This is essentially due to the assumption that V is of $(-, -]$-type.)

♠ In the spirit of [34],

$$\psi(= 1) : \mathbb{R} \longrightarrow \mathbb{R}_\gamma, \quad \text{where } \mathbb{R} \text{ stands for } \mathbb{R} \text{ with the usual topology,}$$

is a continuous map. Therefore, we can consider $\mathcal{H} := \psi^{-1} \mathcal{G}$. For every open subset (a, b) in $\mathbb{R}$,

$$\mathcal{H}((a, b)) = (\psi^{-1} \mathcal{G})((a, b)) = \mathcal{G}((a, \infty)) = \mathcal{F}((a, \infty)). \tag{A.1}$$

The stalk at a is

$$\mathcal{H}_a = (\psi^{-1} \mathcal{G})_a = \mathcal{G}_{\psi(a)} = \mathcal{G}_a (= V_a) \quad \text{by Exercise A.2.} \tag{A.2}$$

Now, for any persistence k-module $V \subset \mathcal{P}$, define

$$\Phi(V) := \mathcal{H}, \quad \text{that is,} \quad \Phi(V)((a, b)) = \varprojlim_{n \to \infty} V_{a + \frac{1}{n}}. \tag{A.3}$$

Remark A.1 Formulas (A.1) and (A.2) reveal several important facts. First, note that the section of $\mathcal{H}$ over (a, b) is independent of the right endpoint b. In other words, $\mathcal{H}$ can propagate all the way to $+\infty$, which implies that (by the definition of the singular support), $SS(\mathcal{H}) \subset \{\tau \leq 0\}$. At the same time, $\mathcal{H}$ is certainly constructible, since away from the spectrum of V it is a locally constant sheaf. Then combining the decomposition theorem of constructible sheaves and the well-known fact that $SS(\bigoplus F_i) \subset \bigcup SS(\mathcal{F}_i)$, we conclude that $\mathcal{H} \in \text{Sh}_c(\mathbb{R})$.

Remark A.2 Equation (A.3) enables us to define the image of structure maps under Φ, that is,

$$\Phi(\iota_{a,b}) := \varprojlim_{n \to \infty} \iota_{a+\frac{1}{n}, b+\frac{1}{n}}. \tag{A.4}$$

When a, b are not in the spectrum of V, $\Phi(\iota_{a,b}) = \mathrm{res}_{a,b} : \Phi(V)((a, \infty)) \to \Phi(V)((b, \infty))$, the restriction maps of the sheaf $\Phi(V)$.

Example A.1 Let $V = \mathbb{I}_{(1,2]}$, then $\mathcal{F}$ defined in the step ♣ satisfies

$$\mathcal{F}([a, \infty)) = \begin{cases} \mathbf{k}, & \text{if } 1 < a \le 2, \\ 0, & \text{otherwise}, \end{cases} \quad \text{and} \quad \mathcal{F}((a, \infty)) = \begin{cases} \mathbf{k}, & \text{if } 1 \le a < 2, \\ 0, & \text{otherwise}. \end{cases}$$

By a simple computation (or Exercise A.2), for $\mathcal{G}$ which is defined in the step ♡, we know that

$$\mathcal{G}_a = \begin{cases} \mathbf{k}, & \text{if } 1 < a \le 2, \\ 0, & \text{if otherwise}. \end{cases}$$

Therefore, $\Phi(\mathbb{I}_{(1,2]})(= \mathcal{H}) = \mathbf{k}_{(1,2]}$.

Finally, if $f : V \to W$ is a morphism between two persistence $\mathbf{k}$-modules, then $\Phi(f)$ is defined by means of the diagram

$$
\begin{array}{ccc}
V_a \xrightarrow{f_a} W_a \\
\iota_{a,b}^V \downarrow \quad\quad \downarrow \iota_{a,b}^W \\
V_b \xrightarrow{f_b} W_b
\end{array}
\quad \Longrightarrow \quad
\begin{array}{ccc}
\Phi(V)((a, \infty)) \xrightarrow{\Phi(f_a)} \Phi(W)((a, \infty)) \\
\Phi(\iota_{a,b}^V) \downarrow \quad\quad\quad\quad \downarrow \Phi(\iota_{a,b}^W) \\
\Phi(V)((b, \infty)) \xrightarrow{\Phi(f_b)} \Phi(W)((b, \infty))
\end{array}
$$

where

$$\Phi(f_a) = \varprojlim_{n \to \infty} f_{a+\frac{1}{n}}.$$

By the commutativity of the left diagram above,

$$\Phi(\iota_{a,b}^W) \circ \Phi(f_a) = \varprojlim_{n \to \infty} \left(\iota_{a+\frac{1}{n}, b+\frac{1}{n}}^W \circ f_{a+\frac{1}{n}} \right)$$

$$= \varprojlim_{n \to \infty} \left(f_{b+\frac{1}{n}} \circ \iota_{a+\frac{1}{n}, b+\frac{1}{n}}^V \right) = \Phi(f_b) \circ \Phi(\iota_{a,b}^V).$$

To construct Ψ (expected to be the inverse of Φ), we just reverse the order of the construction of Φ, that is, $\spadesuit \xrightarrow{\psi_*} \heartsuit \xrightarrow{\phi^{-1}} \clubsuit$. Explicitly, for any $\mathcal{F} \in \mathrm{Sh}_c(\mathbb{R})$,

$$\Psi(\mathcal{F})_a = \varinjlim_{\epsilon \to 0} \mathcal{F}((a - \epsilon, \infty))(\simeq \mathcal{F}_a). \tag{A.5}$$

Example A.2 Let $\mathcal{F} = \mathbf{k}_{(0,1]}$. By a simple computation,

$$\Psi(\mathcal{F})_a = \varinjlim_{\epsilon \to 0} \mathcal{F}((a - \epsilon, \infty)) = \begin{cases} \mathbf{k}, & \text{if for } 0 < a \leq 1, \\ 0, & \text{if for otherwise,} \end{cases}$$

that is, $\Psi(\mathcal{F}) = \mathbb{I}_{(0,1]}$.

Remark A.3 Note that (A.2) and (A.5) together imply that $\Psi \circ \Phi$ and $\Phi \circ \Psi$ are identities. Details are left as an exercise.

A.2 Computations of $R\mathcal{H}om$

Since the computational results in the form of $R\mathcal{H}om(\mathcal{F}, \mathcal{G})$ or $R\mathrm{Hom}(\mathcal{F}, \mathcal{G})$ appear quite often in the main body of this book, we provide, for the reader's convenience, a detailed computation in the case where $\mathcal{F} = \mathbf{k}_I$ and $\mathcal{G} = \mathbf{k}_J$, where I and J are intervals in $\mathbb{R}$. It turns out the answers are very sensitive to their relative positions. We will start from the following basic result.

Theorem A.2 *Fix $a, b \in \mathbb{R}$.*

$$R\mathcal{H}om(\mathbf{k}_{[a,b)}, \mathbf{k}_{[c,\infty)}) = \begin{cases} 0, & \text{for } c \geq b, \\ \mathbf{k}_{[c,b]}, & \text{for } a \leq c < b, \\ \mathbf{k}_{(a,b]}, & \text{for } c < a. \end{cases}$$

Before proving this theorem, we need some preparation.[1] Denote by $\mathrm{Sh}(\mathbb{R})$ the category of sheaves of $\mathbf{k}$-modules over $\mathbb{R}$.

Lemma A.1 *Let A and B be two intervals of $\mathbb{R}$ and let A be closed. Suppose $\mathcal{F} \in \mathrm{Sh}(\mathbb{R})$ is such that $\mathrm{supp}(\mathcal{F}) \subset A$. Then*

(1) *if $A \subset B$, then $R\Gamma_B \mathcal{F} \simeq \mathcal{F}$;*
(2) *if $A \cap \bar{B} = \emptyset$, then $R\Gamma_B \mathcal{F} \simeq 0$.*

---

[1] This is based on a discussion with Semyon Alesker.

Proof Consider the inclusion $i : A \hookrightarrow \mathbb{R}$. We know that $i_* i^{-1} \mathcal{F} = \mathcal{F}$. Take an injective resolution of $i^{-1} \mathcal{F}$, that is

$$0 \longrightarrow i^{-1} \mathcal{F} \longrightarrow I^1 \longrightarrow I^2 \longrightarrow \cdots$$

Applying i_*, which is exact because A is closed, we get an exact sequence,

$$0 \longrightarrow \mathcal{F} \longrightarrow i_* I^1 \longrightarrow i_* I^2 \longrightarrow \cdots$$

where, for any $n \geq 1$, $i_* I^n$ is still injective because the functor $\mathcal{F} \to$ $\mathrm{Hom}(\mathcal{F}, i_* I^n) = \mathrm{Hom}(i^{-1} \mathcal{F}, I^n)$ is exact (since I is injective). Therefore, $(i_* I^{\bullet})$ is an injective resolution of $\mathcal{F}$. Moreover, for any $n \geq 1$, $\mathrm{supp}(i_* I^n) \subset A$. Then

$$R\Gamma_B \mathcal{F} = 0 \longrightarrow \Gamma_B(i_* I^1) \longrightarrow \Gamma_B(i_* I^2) \longrightarrow \cdots$$

By the definition of Γ_B (see Definition 2.3.8 in [32]),

(1) if $A \subset B$, then $\Gamma_B(i_* I^n) = i_* I^n$;
(2) if $A \cap \bar{B} = \emptyset$, then $\Gamma_B(i_* I^n) = 0$;

for any $n \geq 1$. So, up to quasi-isomorphisms, we get the desired conclusion. $\qquad\square$

Remark A.4 The proof of Lemma A.1 did not use any specific property of $\mathbb{R}$, therefore, the same conclusion works for any (smooth) manifold X.

Example A.3 Fix two numbers $x, y \in \mathbb{R}$. Then

(1) If $x \leq y$, then $R\Gamma_{[x,\infty)} \mathbf{k}_{[y,\infty)} = \mathbf{k}_{[y,\infty)}$.
(2) If $x > y$, then $R\Gamma_{[x,\infty)} \mathbf{k}_{(-\infty,y)} = 0$.
(3) If $z > y > x$, then $R\Gamma_{[x,y)} \mathbf{k}_{[z,\infty)} = 0$.

Proposition A.1 $R\Gamma_{[x,\infty)} (\mathbf{k}_{[y,\infty)}) \simeq \mathbf{k}_{(x,\infty)}$ *if* $x > y$.

Proof Consider the short exact sequence

$$0 \longrightarrow \mathbf{k}_{(-\infty,y)} \longrightarrow \mathbf{k}_{\mathbb{R}} \longrightarrow \mathbf{k}_{[y,\infty)} \longrightarrow 0$$

Applying $R\Gamma_{[x,\infty)}$, we get the distinguished triangle

$$R\Gamma_{[x,\infty)} \mathbf{k}_{(-\infty,y)} \longrightarrow R\Gamma_{[x,\infty)} \mathbf{k}_{\mathbb{R}} \longrightarrow R\Gamma_{[x,\infty)} \mathbf{k}_{[y,\infty)} \xrightarrow{+1}$$

By Example A.3 (2), $R\Gamma_{[x,\infty)} \mathbf{k}_{(-\infty,y)} = 0$, and then

$$R\Gamma_{[x,\infty)} \mathbf{k}_{\mathbb{R}} \simeq R\Gamma_{[x,\infty)} \mathbf{k}_{[y,\infty)}. \tag{A.6}$$

Now, by (iv) in Proposition 2.4.6 (or (2.6.32)) in [32], we have another useful distinguished triangle, as follows:

$$R\Gamma_{[x,\infty)}\mathbf{k}_\mathbb{R} \longrightarrow R\Gamma_\mathbb{R}\mathbf{k}_\mathbb{R} \longrightarrow R\Gamma_{(-\infty,x)}\mathbf{k}_\mathbb{R} \xrightarrow{+1} \qquad\qquad \text{(A.7)}$$

Note that $R\Gamma_\mathbb{R}\mathbf{k}_\mathbb{R} = \mathbf{k}_\mathbb{R}$. Moreover, $\Gamma_{(-\infty,x)}\mathbf{k}_\mathbb{R} = (j_* \circ j^{-1})(\mathbf{k}_\mathbb{R})$, where j : $(-\infty, x) \hookrightarrow \mathbb{R}$ by item (iii) in Proposition 2.3.9 in [32]. Therefore,

$$R\Gamma_{(-\infty,x)}\mathbf{k}_\mathbb{R} = (Rj_* \circ j^{-1})(\mathbf{k}_\mathbb{R}) = Rj_*(\mathbf{k}(-\infty,x)) = \mathbf{k}_{(-\infty,x]},$$

where $\mathbf{k}(-\infty, x)$ is the constant sheaf over $(-\infty, x)$ and the final step is checked at stalks. Therefore (A.7) reduces to

$$R\Gamma_{[x,\infty)}\mathbf{k}_\mathbb{R} \longrightarrow \mathbf{k}_\mathbb{R} \xrightarrow{res} \mathbf{k}_{(-\infty,x]} \xrightarrow{+1}$$

and, up to quasi-isomorphism, we have $R\Gamma_{[x,\infty)}\mathbf{k}_\mathbb{R} = \mathbf{k}_{(x,\infty)}$. Now (A.6) completes the proof.   □

We are ready to compute $R\mathcal{H}om(\mathbf{k}_{[a,b)}, \mathbf{k}_{[c,\infty)}) = R\Gamma_{[a,b)}\mathbf{k}_{[c,\infty)}$ (by (2.3.16) in Proposition 2.3.10 in [32]).

Proof (*of Theorem A.2*) We will carry out the computations case by case. Here, we always assume $b > a$.

 (i) When $c > b$, by (3) in Example A.3, we know that $R\Gamma_{[a,b)}\mathbf{k}_{[c,\infty)} = 0$.
 (ii) When $c = b$, by (2.6.32) in [32], consider the distinguished triangle

$$R\Gamma_{[b,\infty)}\mathbf{k}_{[b,\infty)} \longrightarrow R\Gamma_{[a,\infty)}\mathbf{k}_{[b,\infty)} \longrightarrow R\Gamma_{[a,b)}\mathbf{k}_{[b,\infty)} \xrightarrow{+1}$$

By Example A.3 (1), the first and the second terms are both equal to $\mathbf{k}_{[b,\infty)}$. So we get

$$\mathbf{k}_{[b,\infty)} \longrightarrow \mathbf{k}_{[b,\infty)} \xrightarrow{res} R\Gamma_{[a,b)}\mathbf{k}_{[b,\infty)} \xrightarrow{+1}$$

which implies that $R\Gamma_{[a,b)}\mathbf{k}_{[b,\infty)} = 0$.
(iii) When $a < c < b$, again consider the following distinguished triangle

$$R\Gamma_{[b,\infty)}\mathbf{k}_{[c,\infty)} \longrightarrow R\Gamma_{[a,\infty)}\mathbf{k}_{[c,\infty)} \longrightarrow R\Gamma_{[a,b)}\mathbf{k}_{[c,\infty)} \xrightarrow{+1}$$

By Proposition A.1, $R\Gamma_{[b,\infty)}\mathbf{k}_{[c,\infty)} = \mathbf{k}_{(b,\infty)}$. By Example A.3 (1), $R\Gamma_{[a,\infty)}\mathbf{k}_{[c,\infty)} = \mathbf{k}_{[c,\infty)}$. So we get

$$\mathbf{k}_{(b,\infty)} \longrightarrow \mathbf{k}_{[c,\infty)} \xrightarrow{res} R\Gamma_{[a,b)}\mathbf{k}_{[c,\infty)} \xrightarrow{+1}$$

which implies that $R\Gamma_{[a,b)}\mathbf{k}_{[c,\infty)} = \mathbf{k}_{[c,b]}$.

(iv) When $c = a$, the same argument as in (iii) implies that $R\Gamma_{[a,b)}\mathbf{k}_{[a,\infty)} = \mathbf{k}_{[a,b]}$.

(v) When $c < a$, again consider the distinguished triangle

$$R\Gamma_{[b,\infty)}\mathbf{k}_{[c,\infty)} \longrightarrow R\Gamma_{[a,\infty)}\mathbf{k}_{[c,\infty)} \longrightarrow R\Gamma_{[a,b)}\mathbf{k}_{[c,\infty)} \xrightarrow{+1}$$

By Proposition A.1, $R\Gamma_{[b,\infty)}\mathbf{k}_{[c,\infty)} = \mathbf{k}_{(b,\infty)}$ and $R\Gamma_{[a,\infty)}\mathbf{k}_{[c,\infty)} = \mathbf{k}_{(a,\infty)}$. So by the exact sequence (v) in Proposition 2.16 in [32], we get

$$\mathbf{k}_{(b,\infty)} \longrightarrow \mathbf{k}_{(a,\infty)} \xrightarrow{res} R\Gamma_{[a,b)}\mathbf{k}_{[c,\infty)} \xrightarrow{+1}$$

which implies that $R\Gamma_{[a,b)}\mathbf{k}_{[c,\infty)} = \mathbf{k}_{(a,b]}$.

$\square$

Corollary A.2 *Let $a < b$ and $c < d$ in $\mathbb{R}$. Then*

$$R\mathcal{H}om(\mathbf{k}_{[a,b)}, \mathbf{k}_{[c,d)}) = \begin{cases} R\mathcal{H}om(\mathbf{k}_{[a,b)}, \mathbf{k}_{[c,\infty)}) & \text{for } d \geq b, \\ \mathbf{k}_{[c,d)}, & \text{for } a \leq c < d < b, \\ \mathbf{k}_{(a,d)}, & \text{for } c < a < d < b, \\ \mathbf{k}_{\{a\}}[-1], & \text{for } d = a, \\ 0, & \text{for } d < a, \end{cases}$$

Proof We will keep using the following distinguished triangle

$$R\Gamma_{[a,b)}\mathbf{k}_{[c,d)} \longrightarrow R\Gamma_{[a,b)}\mathbf{k}_{[c,\infty)} \longrightarrow R\Gamma_{[a,b)}\mathbf{k}_{[d,\infty)} \xrightarrow{+1}$$

- When $d \geq b$, Theorem A.2 shows that the third term is 0, which implies that $R\Gamma_{[a,b)}\mathbf{k}_{[c,d)} \simeq R\Gamma_{[a,b)}\mathbf{k}_{[c,\infty)}$.
- When $a \leq c < d < b$, Theorem A.2, shows that

$$R\Gamma_{[a,b)}\mathbf{k}_{[c,d)} \longrightarrow \mathbf{k}_{[c,b]} \longrightarrow \mathbf{k}_{[d,b]} \xrightarrow{+1}$$

which implies that $R\Gamma_{[a,b)}\mathbf{k}_{[c,d)} = \mathbf{k}_{[c,d)}$.

- When $c < a < d < b$, Theorem A.2, shows that

$$R\Gamma_{[a,b)}\mathbf{k}_{[c,d)} \longrightarrow \mathbf{k}_{(a,b]} \longrightarrow \mathbf{k}_{[d,b]} \xrightarrow{+1}$$

which implies that $R\Gamma_{[a,b)}\mathbf{k}_{[c,d)} = \mathbf{k}_{(a,d)}$.

- When $d = a$, Theorem A.2 gives

$$R\Gamma_{[a,b)}\mathbf{k}_{[c,d)} \longrightarrow \mathbf{k}_{(a,b]} \xrightarrow{*} \mathbf{k}_{[a,b]} \xrightarrow{+1}.$$

Since $[a, b] \supset (a, b]$, the map $*$ above will change from restriction to inclusion. Therefore, the non-trivial term will appear in the degree-1 term, that is,

$$0 \longrightarrow 0 \longrightarrow \mathbf{k}_{(a,b]} \longrightarrow \mathbf{k}_{[a,b]} \longrightarrow \mathbf{k}_{\{a\}} \longrightarrow 0 \longrightarrow 0 \longrightarrow \cdots$$

which implies that $R^1\Gamma_{[a,b)}\mathbf{k}_{(c,d)} = \mathbf{k}_{\{a\}}$, and is equal to 0 in other degrees.

- When $d < a$, Theorem A.2 gives

$$R\Gamma_{[a,b)}\mathbf{k}_{(c,d)} \longrightarrow \mathbf{k}_{(a,b]} \overset{*}{\longrightarrow} \mathbf{k}_{(a,b]} \overset{+1}{\longrightarrow}$$

which implies that $R\Gamma_{[a,b)}\mathbf{k}_{(c,d)} = 0$.

$\square$

Remark A.5   In Corollary A.2, we have seen that there exist examples where $R\mathcal{H}om$ can have non-trivial degree-1 terms. Here we want to address the strong claim that for any two sheaves $\mathcal{F}, \mathcal{G} \in \mathrm{Sh}(\mathbb{R})$, for any $j \geq 2$, $R^j\mathcal{H}om(\mathcal{F}, \mathcal{G}) = 0$. This comes from the fact that the homological dimension of $\mathrm{Sh}(\mathbb{R})$ is 2 (see Theorem 5.11 in [5] for a general result). Therefore, our claim follows by replacing $\mathcal{G}$ with an injective resolution $0 \to I^1 \to I^2 \to 0 \to \cdots$ with length at most 2 and applying the functor $R\mathcal{H}om(\mathcal{F}, \cdot)$.

In a different direction, instead of $R\mathcal{H}om(\mathbf{k}_{[a,b)}, \mathbf{k}_{(c,d)})$, people are very often interested in the $\mathbf{k}$-module $R\mathrm{Hom}(\mathbf{k}_{[a,b)}, \mathbf{k}_{(c,d)})$. Since we have

$$R\mathrm{Hom}(\mathbf{k}_{[a,b)}, \mathbf{k}_{(c,d)}) = R\Gamma(\mathbb{R}, -) \circ R\mathcal{H}om(\mathbf{k}_{[a,b)}, \mathbf{k}_{(c,d)}), \tag{A.8}$$

Corollary A.2 implies Theorem 3.2.

A.3   Dynamics of the GKS Sheaf Quantization

In A.3 in [26], an algebraic trick was introduced, lifting from T^*M to $T^*(M \times \mathbb{R})$ (and adjusting some 0-section part), in order to fit the homogenous machinery developed in [26]. In this section, we will give a pure dynamical explanation of this trick.[2]

Dynamical motivation of homogenization. The motivation comes from the attempt to include the Plank constant $\hbar$ in classical mechanics. This philosophy is frequently encountered in *semi-classical analysis*. For instance, for Lagrangian mechanics, instead of a Lagrangian $L_a(m, \dot{m}) : \mathbb{R} \times TM \to \mathbb{R}$, where $a \in \mathbb{R}$ is the time parameter, one considers a rescaled Lagrangian $\frac{1}{\hbar}L_a$. By the well-known

---

[2]This is based on a discussion with Alejandro Uribe. This dynamical explanation is also elaborated in [41].

transformation from Lagrangian mechanics to Hamiltonian mechanics, that is, $H_a(m, p) = p\dot{m} - L_a(m, \dot{m})\big|_{p=\frac{\partial L}{\partial \dot{m}}}$, introducing the variable $\xi = \frac{p}{\hbar}$ results in the following computation:

$$(H_\hbar)_a(m, \xi) = \xi\dot{m} - \left(\frac{1}{\hbar}L_a\right)(m, \dot{m})\Big|_{\xi=\frac{\partial L/\hbar}{\partial \dot{m}}}$$

$$= \frac{p}{\hbar}\dot{m} - \frac{1}{\hbar}L_a(m, \dot{m})\Big|_{p=\frac{\partial L}{\partial \dot{m}}}$$

$$= \frac{1}{\hbar}H_a(m, p) = \frac{1}{\hbar}H_a(m, \xi\hbar).$$

In other words, denoting $\tau = \frac{1}{\hbar} \in \mathbb{R}_{>0}$, we get the function $(H_{1/\tau})_a(m, \xi) = \tau H_a\left(m, \frac{\xi}{\tau}\right)$. Therefore, let us consider the function

$$\widetilde{H} : \mathbb{R} \times T^*M \times T^*_{>0}\mathbb{R} \longrightarrow \mathbb{R} \quad \text{given by} \quad \widetilde{H}(a, m, \xi, t, \tau) = \tau H_a\left(m, \frac{\xi}{\tau}\right),$$
$$\text{(A.9)}$$

where t is the dual coordinate of τ. The symplectic manifold $(T^*_{\{\tau>0\}}(M \times \mathbb{R}), dm \wedge d\xi + dt \wedge d\tau)$ can be obtained by contactization-then-symplectization of symplectic manifold $(T^*M, dm \wedge dp)$ where $p = \xi/\tau$.

It is crucial that both t and τ are viewed as dynamical variables (so that we can take derivatives with respect to them). Via a standard computation, the using standard symplectic structure on $T^*_{\{\tau>0\}}(M \times \mathbb{R})$ as above, one can get the Hamiltonian system of first-order differential equations. The detailed result is in formulas (14)–(17) in [41]. Here, we just want to emphasize that, for the t-component,

$$\dot{t} = H_a\left(m, \frac{\xi}{\tau}\right) - \frac{\xi}{\tau}\frac{\partial H}{\partial \xi}\left(m, \frac{\xi}{\tau}\right) = H_a(m, p) - p\dot{m}(= -L_a(m, \dot{m})).$$

This fact that the evolution of the t-component is governed by the initial (negative) Lagrangian L_a is remarkable!

Since this $\widetilde{H}$ generates an isotopy of Hamiltonian diffeomorphisms $\Phi_a : \mathbb{R} \times T^*_{\{\tau>0\}}(M \times \mathbb{R}) \to T^*_{\{\tau>0\}}(M \times \mathbb{R})$, we can consider its Lagrangian suspension as a Lagrangian subspace of $T^*\mathbb{R} \times T^*_{\{\tau>0\}}(M \times \mathbb{R}) \times T^*_{\{\tau>0\}}(M \times \mathbb{R})$,

$$\widetilde{\Lambda} := \left\{((a, -\widetilde{H}), x, -\Phi_a(x)) \mid a \in \mathbb{R}, x \in T^*_{\{\tau>0\}}(M \times \mathbb{R})\right\}$$

$$= \left\{\left(a, -\tau H_a\left(m, \frac{\xi}{\tau}\right), (m, \xi, t, \tau), \left(-\tau\phi_a\left(m, \frac{\xi}{\tau}\right)\right), t + (*), -\tau\right) \,\middle|\, \begin{matrix} (m, \xi/\tau) \in T^*M \\ \tau > 0 \end{matrix}\right\},$$

where ϕ_a is the isotopy of Hamiltonian diffeomorphisms generated by H_a on T^*M and (∗) is defined, for any fixed terminal time $A \in \mathbb{R}$, by

$$(*) = \int_0^A (H_a(m, p) - p\dot{m}) \circ \phi_a da := F_A(m, p), \tag{A.10}$$

where this $F_A(m, p)$ is, in the language of symplectic geometry, the standard symplectic action functional.

Geometry of GKS's sheaf quantization (revised). Recall that the main theorem in [26] (Theorem 4.1) says that for any compactly supported Hamiltonian isotopy ϕ_a, generated by H on T^*M, there exists a unique sheaf $\mathcal{K} \in \mathcal{D}(\mathbf{k}_{\mathbb{R} \times M \times \mathbb{R} \times M \times \mathbb{R}})$ (where we extend I to $\mathbb{R}$) such that $SS(\mathcal{K}) \cap (T^*\mathbb{R} \times T^*_{\{\tau>0\}}(M \times \mathbb{R}) \times T^*_{\{\tau>0\}}(M \times \mathbb{R})) \subset \widetilde{\Lambda} \cup \{0\text{-section}\}$, i.e., the geometry of $\mathcal{K}$ is characterized by $\widetilde{\Lambda}$. In order to get a more friendly expression, consider the subtract map $\bar{s} : \mathbb{R} \times \mathbb{R} \to \mathbb{R}$ by $\bar{s}(t_1, t_2) = t_1 - t_2$. Then on the co-vector part, $\bar{s}$ induces the anti-diagonal embedding, $\bar{s}^* : \mathbb{R}^* \to \mathbb{R}^* \times \mathbb{R}^*$ by $\bar{s}^*(\tau) = (\tau, -\tau)$. Therefore, one has $R\bar{s}_*\mathcal{K} \in \mathcal{D}(\mathbf{k}_{\mathbb{R} \times M \times M \times \mathbb{R}})$, and also

$$SS(R\bar{s}_*\mathcal{K}) \cap (T^*\mathbb{R} \times T^*M \times T^*M \times T^*_{>0}\mathbb{R}) \subset \Lambda \cup \{0\text{-section}\},$$

where

$$\Lambda := \left\{ (a, -\tau H_a(m, \xi/\tau), m, \xi, -\tau\phi_a(m, \xi/\tau), -F_a(m, p), \tau) \,\middle|\, \begin{array}{c} (m, \xi/\tau) \in T^*M \\ \tau > 0 \end{array} \right\}. \tag{A.11}$$

Note that along $\{\tau = 1\}$ we have $\xi/\tau = p = \xi/1$, and so the reduction of Λ along $\{\tau = 1\}$ is the submanifold

$$\Lambda_0 = \left\{ (a, -H_a(m, p), m, p, -\phi_a(m, p)) \,\middle|\, (m, \xi) \in T^*M \right\},$$

which is just the Lagrangian suspension of ϕ_a on T^*M. Therefore, $R\bar{s}_*\mathcal{K}$ represents a "semi-classical counterpart" of the usual Lagrangian suspension.

In fact, in the concrete case when $M = \mathbb{R}^n$, we have seen this geometric constraint (A.11) from the direct construction $\mathcal{F}_S := \mathbf{k}_{\{S+t \geq 0\}} \in \mathcal{D}(\mathbf{k}_{\mathbb{R} \times M \times M \times \mathbb{R}})$ (modulo convolution) using the generating function method, where F_a in (A.10) is the same as $S(a, \cdot)$ in (4.21). Such $\mathcal{F}_S$ can also be called a sheaf quantization of Hamiltonian isotopy ϕ_a because $\mathcal{K}$ and $\mathcal{F}_S$ are comparable to some extent (L. Polterovich's comment). In order to compare these sheaves, we need to modify them into the same space. Recall that the singular support of $\mathcal{F}_S$ always has its τ-component in $\{\tau \geq 0\}$. Let us denote $\mathcal{K}_+$ as the restriction of $\mathcal{K}$ on the part where $\{\tau > 0\}$. Then we can consider the following two comparisons.

(i) $R\bar{s}_*\mathcal{K}_+, \mathcal{F}_S \in \mathcal{D}(\mathbf{k}_{\mathbb{R} \times M \times M \times \mathbb{R}})$;
(ii) $\mathcal{K}_+, \bar{s}^{-1}\mathcal{F}_S \in \mathcal{D}(\mathbf{k}_{\mathbb{R} \times M \times \mathbb{R} \times M \times \mathbb{R}})$.

By the pullback formula of SS (see Proposition 2.7), one can check that $\bar{s}^{-1}\mathcal{F}_S$ satisfies $SS(\bar{s}^{-1}\mathcal{F}_S) \subset \tilde{\Lambda} \cup \{0\text{-section}\}$ and $(\bar{s}^{-1}\mathcal{F}_S)|_{a=0} = \mathbf{k}_\Delta$. Then by the uniqueness part of the GKS sheaf quantization (which also holds for the $\{\tau \geq 0\}$-restriction), $\mathcal{K}_+ \simeq \bar{s}^{-1}\mathcal{F}_S$ in the case (ii). On the other hand, the interested reader can check that the method that establishes the uniqueness of sheaf quantization works perfectly well for the space $\mathbb{R} \times M \times M \times \mathbb{R}$ (instead of $\mathbb{R} \times (M \times \mathbb{R}) \times (M \times \mathbb{R})$), since the constraint of singular support as demonstrated in Proposition 4.1 is only from the first $\mathbb{R}$-component (the variable of time). Here, only the last two $\mathbb{R}$-components are changed (to be $\mathbb{R}$) by the map $\bar{s}$. Therefore, this uniqueness shows that $R\bar{s}_* \mathcal{K}_+ \simeq \mathcal{F}_S$ in the case (i) as well.

References

1. Asano, T., Ike, Y.: Persistence-like distance on Tamarkin's category and symplectic displacement energy (2017). Preprint. arXiv: 1712.06847
2. Audin, M., Damian, M.: Morse Theory and Floer Homology. Springer, London (2014)
3. Bauer, U., Lesnick, M.: Induced matchings of barcodes and the algebraic stability of persistence. In: Proceedings of the Thirtieth Annual Symposium on Computational Geometry, p. 355. ACM, New York (2014)
4. Biran, P., Polterovich, L., Salamon, D.: Propagation in Hamiltonian dynamics and relative symplectic homology. Duke Math. J. **119**(1), 65–118 (2003)
5. Björk, J.: Analytic $\mathcal{D}$-modules and Applications. Mathematics and its Applications, vol. 247. Kluwer Academic Publishers Group, Dordrecht (1993)
6. Bubenik, P., de Silva, V., Nanda., V.: Higher interpolation and extension for persistence modules. SIAM J. Appl. Algebra Geom. **1**(1), 272–284 (2017)
7. Chaperon, M.: Phases génératrices en géométrie symplectique. In: Les rencontres physiciens-mathématiciens de Strasbourg-RCP25, vol. 41, pp. 191–197 (1990)
8. Chazal, F., Cohen-Steiner, D., Glisse, M., Guibas, L.J., Oudot, S.: Proximity of persistence modules and their diagrams. In: Proceedings of the Twenty-fifth Annual Symposium on Computational Geometry, pp. 237–246. ACM, New York (2009)
9. Chazal, F., de Silva, V., Glisse, M., Oudot, S.: The Structure and Stability of Persistence Modules. SpringerBriefs in Mathematics. Springer, Cham (2016)
10. Chiu, S.F.: Nonsqueezing property of contact balls. Duke Math. J. **166**(4), 605–655 (2017)
11. Cohen-Steiner, D., Edelsbrunner, H., Harer, J.: Stability of persistence diagrams. Discrete Comput. Geom. **37**(1), 103–120 (2007)
12. Crawley-Boevery, W.: Decomposition of pointwise finite-dimensional persistence modules. J. Algebra Appl. **14**(05), 1550066 (2015)
13. Curry, J.: Sheaves, cosheaves and applications (2013). Preprint. arXiv: 1303.3255
14. D'Agnolo, A., Kashiwara, M.: Riemann-Hilbert correspondence for holonomic $\mathcal{D}$-modules. Publications mathématiques de l'IHÉS **123**(1), 69–197 (2016)
15. Eliashberg, Y., Kim, S.S., Polterovich, L.: Geometry of contact transformations and domains: orderability versus squeezing. Geom. Topol. **10**(3), 1635–1747 (2006)
16. Floer, A.: Symplectic fixed points and holomorphic spheres. Commun. Math. Phys. **120**(4), 575–611 (1989)
17. Floer, A., Hofer, H.: Symplectic homology I open sets in $\mathbb{C}^n$. Math. Z. **215**(1), 37–88 (1994)
18. Fraser, M.: Contact non-squeezing at large scale in $\mathbb{R}^{2n} \times S^1$. Int. J. Math. **27**(13), 1650107 (2016)

© Springer Nature Switzerland AG 2020
J. Zhang, *Quantitative Tamarkin Theory*, CRM Short Courses,
https://doi.org/10.1007/978-3-030-37888-2

19. Frauenfelder, U., Ginzburg, V., Schlenk, F.: Energy-capacity inequalities via an action selector. In: Geometry, Spectral Theory, Groups, and Dynamics. Contemporary Mathematics, vol. 387, pp. 129–152. AMS, Providence (2005)
20. Gabriel, P.: Unzerlegbare darstellungen I. Manuscripta Math. **6**(1), 71–103 (1972)
21. Gelfand, S., Manin, Y.: Methods of Homological Algebra. Springer Monographs in Mathematics, 2nd edn. Springer, Berlin (2003)
22. Gromov, M.: Pseudo holomorphic curves in symplectic manifolds. Invent. Math. **82**(2), 307–347 (1985)
23. Guillermou, S.: The Gromov-Eliashberg theorem by microlocal sheaf theory (2013). Preprint. arXiv: 1311.0187
24. Guillermou, S.: The three cusps conjecture (2016). Preprint. arXiv: 1603.07876
25. Guillermou, S.: Sheaves and symplectic geometry of cotangent bundles (2019). Preprint. arXiv: 1905.07341
26. Guillermou, S., Kashiwara, M., Schapira, P.: Sheaf quantization of Hamiltonian isotopies and applications to nondisplaceability problems. Duke Math. J. **161**(2), 201–245 (2012)
27. Guillermou, S., Schapira, P.: Microlocal theory of sheaves and Tamarkin's non displaceability theorem. In: Homological Mirror Symmetry and Tropical Geometry, pp. 43–85. Springer, Cham (2014)
28. Hartshorne, R.: Algebraic Geometry. Graduate Texts in Mathematics, No. 52. Springer, New York (1977)
29. Hofer, H., Zehnder, E.: Symplectic Invariants and Hamiltonian Dynamics. Birkhäuser, Basel (1994)
30. Huybrechts, D.: Fourier-Mukai Transforms in Algebraic Geometry. Oxford University Press on Demand, Oxford (2006)
31. Kashiwara, M.: The Riemann-Hilbert problem for holonomic systems. Publ. Res. Inst. Math. Sci. **20**(2), 319–365 (1984)
32. Kashiwara, M., Schapira, P.: Sheaves on Manifolds. Grundlehren der Mathematischen Wissenschaften, vol. 292. Springer, Berlin (1990). With a chapter in French by Christian Houzel
33. Kashiwara, M., Schapira, P.: Categories and Sheaves. Grundlehren der mathematischen Wissenschaften, vol. 332. Springer, Berlin (2006)
34. Kashiwara, M., Schapira, P.: Persistent homology and microlocal sheaf theory. J. Appl. Comput. Topol. **2**(1–2), 83–113 (2018)
35. Lalonde, F., McDuff, D.: The geometry of symplectic energy. Ann. Math. **141**, 349–371 (1995)
36. McDuff, D., Salamon, D.: Introduction to Symplectic Topology, 2nd edn. Oxford University Press, Oxford (1998)
37. McDuff, D., Salamon, D.: J-holomorphic Curves and Symplectic Topology, vol. 52. American Mathematical Society, Providence (2012)
38. Nadler, D., Zaslow, E.: Constructible sheaves and the Fukaya category. J. Am. Math. Soc. **22**(1), 233–286 (2009)
39. Ng, L., Rutherford, D., Shende, V., Sivek, S., Zaslow, E.: Augmentations are sheaves (2015). Preprint. arXiv: 1502.04939
40. Oancea, A.: A survey of Floer homology for manifolds with contact type boundary or symplectic homology. In: Symplectic geometry and Floer homology. A survey of Floer homology for manifolds with contact type boundary or symplectic homology. Ensaios Matemáticos, vol. 7, pp. 51–91. Sociedade Brasileira de Matemática, Rio de Janeiro (2004)
41. Paul, T., Uribe, A.: The semi-classical trace formula and propagation of wave packets. J. Funct. Anal. **132**(1), 192–249 (1995)
42. Polterovich, L.: Symplectic displacement energy for Lagrangian submanifolds. Ergodic Theory Dynam. Syst. **13**(2), 357–367 (1993)
43. Polterovich, L., Rosen, D., Samvelyan, K., Zhang, J.: Topological Persistence in Geometry and Analysis (2019). Preprint. arXiv: 1904.04044
44. Polterovich, L., Shelukhin, E.: Autonomous Hamiltonian flows, Hofer's geometry and persistence modules. Selecta Math. (N.S.) **22**(1), 227–296 (2016)

45. Polterovich, L., Shelukhin, E., Stojisavljević, V.: Persistence modules with operators in Morse and Floer theory. Moscow Math. J. **17**(4), 757–786 (2017)
46. Sabloff, J.M., Traynor, L.: Obstructions to Lagrangian cobordisms between Legendrians via generating families. Algebr. Geom. Topol. **13**(5), 2733–2797 (2013)
47. Salamon, D.: Lectures on Floer homology. In: Symplectic Geometry and Topology (Park City, UT, 1997), vol. 7, pp. 143–229. American Mathematical Society, Providence (1999)
48. Sandon, S.: Contact homology, capacity and non-squeezing in $\mathbb{R}^{2n} \times S^1$ via generating functions. Ann. Inst. Fourier **61**(1), 145–185 (2011)
49. Sandon, S.: Generating functions in symplectic topology. Lecture notes for the CIMPA research school on geometric methods in classical dynamical systems, Santiago (2014)
50. Schlenk, F.: Applications of Hofer's geometry to Hamiltonian dynamics. Comment. Math. Helv. **81**(1), 105–121 (2006)
51. Shende, V., Treumann, D., Zaslow, E.: Legendrian knots and constructible sheaves. Invent. Math. **207**(3), 1031–1133 (2017)
52. Tamarkin, D.: Microlocal condition for non-displaceability. In: Algebraic and Analytic Microlocal Analysis, pp. 99–223. Springer, Cham (2013)
53. Tamarkin, D.: Microlocal category (2015). Preprint. arXiv: 1511.08961
54. Traynor, L.: Symplectic homology via generating functions. Geom. Funct. Anal. **4**(6), 718–748 (1994)
55. Tsygan, B.: A microlocal category associated to a symplectic manifold. In: Algebraic and Analytic Microlocal Analysis, pp. 225–337. Springer, Cham (2013)
56. Usher, M.: The sharp energy-capacity inequality. Commun. Contemp. Math. **12**(03), 457–473 (2010)
57. Usher, M.: Hofer's metrics and boundary depth. Annales scientifiques de l'École Normale Supérieure **46**(1), 57–129 (2013)
58. Usher, M., Zhang, J.: Persistent homology and Floer-Novikov theory. Geom. Topol. **20**(6), 3333–3430 (2016)
59. Viterbo, C.: Symplectic topology as the geometry of generating functions. Math. Ann. **292**(1), 685–710 (1992)
60. Viterbo, C.: Functors and computations in Floer homology with applications, I. Geom. Funct. Anal. GAFA **9**(5), 985–1033 (1999)
61. Viterbo, C.: An introduction to symplectic topology through sheaf theory (2011). Preprint
62. Zhang, J.: p-cyclic persistent homology and Hofer distance. J. Symplectic Geom. **17**(3), 857–927 (2019)
63. Zomorodian, A., Carlsson, G.: Computing persistent homology. Discrete Comput. Geom. **33**(2), 249–274 (2005)

Index

A
Adjoint functor, 25, 42, 61
Adjoint sheaf, 45, 60, 61
Alexandrov topology, 130
Arnold conjecture, 2, 6, 9, 83

B
Barcode, 7, 8, 29, 65, 115
Base change, 24, 56, 90, 120
Bottleneck distance, 31
Boundary depth, 96, 100

C
Cartesian square, 24
Characteristic foliation, 109, 112
Cohomological functor, 72
Coisotropic (involutive), 35, 108
Conormal bundle, 32, 94
Constructible (sheaf), 8, 35, 59, 67, 115, 130
Convolution, 5, 9, 47–49, 82

D
Derived category, 4, 14, 18, 22, 23, 82
Derived functor, 15, 17, 21, 24, 54
Displacement energy, 8, 69
Distinguished triangle, 6, 22, 34, 75, 101, 112, 134
Dual sheaf identity, 85, 89

E
Energy-capacity inequality, 9, 100
Exact functor (left, right), 14, 16, 20, 23
External tensor, 36

F
Filtered complex, 60
Flabby (sheaf), 17, 24
Fourier-Sato transform, 106

G
Generating function, 3, 46, 53, 55, 65, 70, 102, 109
GH-homology, 66
Gromov-Eliashberg Theorem, 71
Gromov's non-squeezing theorem, 2, 6, 126
Grothendieck composition formula, 24

H
Hamiltonian diffeomorphism, 2, 9, 84, 88, 94, 124, 138
Hamiltonian equations, 110, 138
Hofer norm, 8, 95
Homogeneous Hamiltonian isotopy, 9, 81, 83

I
Injective, 15, 16, 21, 134
Interleaving distance, 29, 69
Interval-type k-module, 7, 27, 30
Isometry Theorem, 31

© Springer Nature Switzerland AG 2020
J. Zhang, *Quantitative Tamarkin Theory*, CRM Short Courses,
https://doi.org/10.1007/978-3-030-37888-2

Printed in the United States
by Baker & Taylor Publisher Services

Printed in the United States
by Baker & Taylor Publisher Services